Beginning

ALGEBRA

Skills Practice Workbook

Factoring
$$8x^5 - 12x^3 = 4x^3(2x^2 - 3)$$

Distributing
$$-5x(x - 6) = -5x^2 + 30x$$

FOIL $x(x) + x(-8) + 7(x) + 7(-8)$
$$(x + 7)(x - 8) = x^2 - x - 56$$

Combine Like Terms $-3x + 5x$
$$12 - \boxed{5x} = 4 - \boxed{3x}$$

Isolate the Unknown $8 = 2x$

Chris McMullen, Ph.D. $4 = x$ $\div 2$

Beginning Algebra Skills Practice Workbook
Factoring, Distributing, FOIL, Combine Like Terms, Isolate the Unknown
Chris McMullen, Ph.D.

www.improveyourmathfluency.com
www.monkeyphysicsblog.wordpress.com
www.chrismcmullen.com

Zishka Publishing
ISBN: 978-1-941691-91-5

Mathematics > Prealgebra
Mathematics > Algebra

Contents

Introduction

The goal of this workbook is to help students develop fluency in specific beginning algebra skills. Most algebra students would benefit from additional practice with these skills.

- Combine like terms: This vital skill is used for almost every topic in algebra. It is used to simplify expressions and to solve equations for unknowns.

- Multiply variables with exponents: When x^m multiplies x^n, the exponents are added together to make x^{m+n}. This fundamental concept is applied when two expressions are multiplied. For example, $(3x^2 + 4)(2x^2 - 5)$ involves $x^2 x^2 = x^4$.

- Distribute: According to the distributive property of arithmetic, $a(b + c) = ab + ac$. This property also applies to algebra. For example, $x(x + 3) = x(x) + x(3) = x^2 + 3x$.

- Distribute minus signs: The distributive property applies to negative numbers, too. For example, $-(x - 8) = -1(x) - 1(-8) = -x + 8$. Here, the two minus signs make a plus sign in the second term.

- Factor: When the distributive property is applied in reverse, it is called factoring. For example, $15x^3 + 20x^2 = 5x^2(3x + 4)$. The greatest common factor (GCF) of 15 and 20 equals 5, and the highest power of x that is common to x^3 and x^2 is x^2.

- The FOIL method: This valuable skill allows students to multiply two expression of the form $(6x - 5)(2x + 7)$. FOIL is an abbreviation for "first, outside, inside, last."

- Square of the sum: An expression of the form $(x + y)^2$ equals $x^2 + 2xy + y^2$. Knowing this formula often comes in handy.

- Difference of squares: An expression of the form $(x + y)(x - y)$ equals $x^2 - y^2$. It is often useful to factor expressions of the form $x^2 - y^2$ as $(x + y)(x - y)$.

- Isolate the unknown: A fundamental way to solve a large class of algebraic equations is to first combine like terms and then isolate the unknown by applying one operation at a time to both sides of the equation.

- Factor quadratic expressions: Some expressions of the form $x^2 + bx + c$ can be put in the form $(x + d)(x + e)$ without using irrational numbers or fractions. Knowing how to do this can save valuable time.

- Answer key. Practice makes permanent, but not necessarily perfect. Check the answers at the back of the book and strive to learn from any mistakes. This will help to ensure that practice makes perfect.

Combine Like Terms

A **variable** is a letter like x or y that represents an unknown number. The reason for calling it a variable is that it equals different values in different problems. For example, x might turn out to equal 3 in one problem and x might turn out to equal 8 in another problem. The value of x varies from problem to problem.

An **equation** has an equal sign, whereas an **expression** does NOT have an equal sign. For example, $6x - 9 = 15$ is an equation, whereas $2x^2 - 3x + 5$ is an expression. This book begins by working with expressions. Later chapters will work with equations.

A **coefficient** is a number that multiplies a variable. For example, 6 is the coefficient in the expression $6x^2$ while 4 is the coefficient in $4x$.

The **terms** of an expression are separated by plus and minus signs. For example, the expression $4x^3 + 6x^2 - 3x + 5$ has four terms, starting with $4x^3$ and ending with 5.

When two terms have the same variable and the same exponent, like $5x^2$ and $3x^2$, the terms are considered to be **like terms**. Two **constant** terms like 7 and 4 (which do NOT have any variable) are also considered to be **like terms**. If the exponents are different, the terms are **unlike**. For example, $2x^2$ and $6x$ are unlike terms because in one case the variable is squared, but in the other case the variable is not squared.

If two terms are **like terms**, they can be combined together by adding or subtracting the coefficients. For example, $5x^2 + 3x^2 = 8x^2$. This only works for like terms.

Chapter 1 Examples

Example 1. Simplify $6x + 9 + 3x - 5$.

Identify the like terms:
$$6x + 9 + 3x - 5 = (6x + 3x) + (9 - 5) = 9x + 4$$
Note: It is not really necessary to write the middle step above that has parentheses. We just wrote this step to help show which terms are like terms. The solution could be written more concisely as $6x + 9 + 3x - 5 = 9x + 4$.

Example 2. Simplify $8x^2 - 6x - 3x^2 + 4x$.

Identify the like terms:
$$8x^2 - 6x - 3x^2 + 4x = (8x^2 - 3x^2) + (-6x + 4x) = 5x^2 - 2x$$
Note: $-6 + 4 = -2$ because -6 is farther from zero than 4.

Example 3. Simplify $x^4 + 5 - 3x^4$.

Identify the like terms:
$$x^4 + 5 - 3x^4 = (x^4 - 3x^4) + 5 = -2x^4 + 5$$
Notes: The constant term, 5, does not have any like terms to combine with. When no coefficient is present before a variable, as in x^4, the coefficient equals 1. For example, $x^4 = 1x^4$. Therefore, $x^4 - 3x^4 = 1x^4 - 3x^4 = -2x^4$. Note that $1 - 3 = -2$ because -3 is farther from zero than 1.

Example 4. Simplify $8x - 3 + 4x - 5 - 5x - 7$.

Identify the like terms:
$$8x - 3 + 4x - 5 - 5x - 7 = (8x + 4x - 5x) + (-3 - 5 - 7) = 7x - 15$$

Example 5. Simplify $4x + 3x^2 - 6 + x^2 - 3x - 2$.

Identify the like terms:
$$4x + 3x^2 - 6 + x^2 - 3x - 2 = (3x^2 + x^2) + (4x - 3x) + (-6 - 2) = 4x^2 + x - 8$$
Notes: $4x - 3x = 1x = x$. When the coefficient of a variable is 1, the coefficient is not written. It is customary to order the terms from highest power to lowest power.

Chapter 1 Problems

Directions: Combine like terms to simplify each expression.

Problem 1. Simplify $4x + 5 + 11x + 3$.

Problem 2. Simplify $2x^2 + 16x + x^2 - 6x$.

Problem 3. Simplify $17x^2 - 10x^2 - 7 + 6$.

Problem 4. Simplify $3x + 9 - 4x$.

Problem 5. Simplify $x^2 + 7 + x^2 + 8$.

Problem 6. Simplify $2x + 36x - 26x$.

Problem 7. Simplify $-12x - 3 + 6x - 14$.

Problem 8. Simplify $-7x^2 + 4x^2 - 6x^2$.

Problem 9. Simplify $27x^2 + 5x^3 + 11x^2 + 8x^3$.

Problem 10. Simplify $6x + 3 + x - 7 + 2x$.

Problem 11. Simplify $13 - 9x + 6$.

Problem 12. Simplify $5x^2 - 8x^2 - 4x^2$.

Problem 13. Simplify $17x^3 - 14x - 12x^3 - 14x$.

Problem 14. Simplify $16 + 13x^2 - 17 - 22x^2 - 6$.

Problem 15. Simplify $25x^3 + 28x^4 - 21x^4 - 15x^3$.

Problem 16. Simplify $-8x - 17 + 4x - 10$.

Problem 17. Simplify $14 + 22x^2 - 19 + 21x^2$.

Problem 18. Simplify $6x - 4x + 9x - x$.

Problem 19. Simplify $25x^5 + 5x^3 + 50x^5 - 6x^3$.

Problem 20. Simplify $3x^2 - 5 - 8x^2 + 4 - 7x^2$.

Problem 21. Simplify $-9x - 2x^3 + 4x + 10x^3$.

Problem 22. Simplify $6x + 12 - 4 - x - 9$.

Problem 23. Simplify $19x^2 + 13x^3 - 2x^2 - 3x^3$.

Problem 24. Simplify $15x + 21 - 12x + 4$.

Problem 25. Simplify $4x^2 - 9x^2 + 13 + 7x^2 + 18$.

Problem 26. Simplify $6x^3 - 5x^5 + 2x^3 - 2x^5$.

Problem 27. Simplify $41x + 10x^2 + 33x + 9x^2 + 8x$.

Problem 28. Simplify $26 - 9x^3 - 26 - 4x^3$.

Problem 29. Simplify $13x^2 + 8 - 14 - 14x^2 + 2$.

Problem 30. Simplify $6x + 12 + 6x - 1 + 3x - 4$.

Problem 31. Simplify $x^3 + 25x^2 - 6x^3 - 22x^2 + 9x^3$.

Problem 32. Simplify $10x - 1 - 4x + 5x - 2 - 3$.

Problem 33. Simplify $-44x + 6x^2 - x + 50x - 7x^2$.

Problem 34. Simplify $15 + 3x^2 - 16 - 2x^2 + 1 - x^2$.

Problem 35. Simplify $6x^3 - 8x^3 - 11x^4 - 16x^4 - 11x^3$.

Problem 36. Simplify $-x - 1 - x - 1 + 3x - 1$.

Problem 37. Simplify $-3x^5 - 4x^5 + 4x^5 - 11 - 3$.

Problem 38. Simplify $21x^2 + 7x + 9x^2 + 14x + 10x$.

Problem 39. Simplify $6x - 12x^3 + 11x^3 - 3x - x^3 + 4x$.

Problem 40. Simplify $x^2 + 2x + 4 + 2x^2 + 2x + 2$.

Problem 41. Simplify $90x + x^4 - 20x - 70x - 23x^4$.

Problem 42. Simplify $x^5 + 8x^2 - x^5 + 17x^2 - 2x^5 - 10x^2$.

Problem 43. Simplify $9 - 4x - 21 + 6x - 11 + 8x$.

Problem 44. Simplify $8 - 7x^2 - 3 + 6x^2 - 1$.

Problem 45. Simplify $6x^3 + 22x - 8x^3 + 11x - 6x^3 + 4x$.

Problem 46. Simplify $-9x^4 - 2x^2 + 4x^4 + 11x^2 - 3x^2$.

Problem 47. Simplify $x - 12 - x + 3x - 4 - 1$.

Problem 48. Simplify $6x^3 - x^2 + 5x^3 + 6x^3 - 3x^2$.

Problem 49. Simplify $-5x^2 - 4x^2 + 8 - 7x^2 + 9 + 10$.

Problem 50. Simplify $16x + 15 + 4x + 8x + 22 + 4$.

Problem 51. Simplify $5 - 11x^3 + x + 6 + 6x^3 + 10$.

Problem 52. Simplify $5x^2 - 6x + 3x - 2x^2 - 7x$.

Problem 53. Simplify $2x^3 - 2x - 14x - 3x^3 - 20x^3 + 14x$.

Problem 54. Simplify $18x + 27x + 5x^2 + 10x^2 + 31x + 22x$.

Problem 55. Simplify $-4x^3 + 4x^2 + 7x + 5x^3 - 5x^2 - 8x$.

Problem 56. Simplify $11x - 4 - 2x + 3x - 2 - 5$.

Problem 57. Simplify $-11x^7 - 11x^4 - 11x^7 - 11x^7 - x^4 - 11x^4$.

Problem 58. Simplify $65x + 46 - x - 12x - 13 + 20$.

Problem 59. Simplify $43x^2 + 4 - 13x^2 + 112 - 12x^2 - 43$.

Problem 60. Simplify $5x^2 - 7x + x^3 - 7x + 2x^3 + 5x^2$.

Problem 61. Simplify $47x^2 - 4 - 6x - x + 11 - 22x^2$.

Problem 62. Simplify $16 - 9x^3 + 4x - 11 - 11x^3 + 2x$.

Problem 63. Simplify $75 - 5x^3 - 45 - x^3 + 3 - 14$.

Problem 64. Simplify $2x^2 + x + 6 + x^2 + 6x + 16$.

Problem 65. Simplify $45x^3 + x^2 - 11x - 25x^3 - 55x^2 - 20x$.

Problem 66. Simplify $28 - 4x + 12 + 6x^2 - 1 + 8x - 4$.

Problem 67. Simplify $-25x^3 - x - 110 - 15x^3 + 15x + 25$.

Problem 68. Simplify $8x^2 - 22 - 31 - 8x^2 - 8x^2 + 4$.

Problem 69. Simplify $20 + 14x^2 - 22 - 3x^3 + 3x^2 - 2x^3$.

Problem 70. Simplify $14 + x - 13x^2 - 5x + 7 - 4x^2$.

Multiply Variables with Exponents

In the expression x^3, the variable x is the **base** and the **exponent** is 3. The exponent is sometimes referred to as a **power**. An exponent indicates repeated multiplication. For example, 2^3 means $2 \times 2 \times 2 = 8$ and $10^4 = 10 \times 10 \times 10 \times 10 = 10,000$.

When x^m multiplies x^n, the exponents get added together: $x^m x^n = x^{m+n}$. Note that $x^m x^n$ means x^m times x^n. For example, if $m = 4$ and $n = 3$, this formula gives $x^4 x^3 = x^7$. You can verify that this rule works by trying different numbers. For example, let $x = 10$, for which $10^4 = 10,000$, $10^3 = 1000$, and $10^7 = 10,000,000$. In this example, $10^4 10^3 = 10^7$ because $(10,000)(1000) = 10,000,000$.

Recall that $x^1 = x$. If no exponent is visible, the exponent equals one.

Chapter 2 Examples (Set 1)

Example 1. Simplify $x^6 x^2$.
Add the exponents: $x^6 x^2 = x^{6+2} = x^8$.

Example 2. Simplify $3x^5 x$.
Note that $x^1 = x$. Add the exponents: $3x^5 x = 3x^5 x^1 = 3x^{5+1} = 3x^6$.

Example 3. Simplify $2x^3 x^2 x^4$.
Add the exponents: $2x^3 x^2 x^4 = 2x^{3+2+4} = 2x^9$.

Chapter 2 Problems (Set 1)

Directions: Apply the rule for multiplying exponents to simplify each expression.

Problem 1. Simplify $x^2 x^4$.

Problem 2. Simplify $x^4 x$.

Problem 3. Simplify $2x^5 x^3$.

Problem 4. Simplify $x^2 x$.

Problem 5. Simplify $3x^4 x^5$.

Problem 6. Simplify $x^3 x^3$.

Problem 7. Simplify $x x^7$.

Problem 8. Simplify $8x^7 x^6$.

Problem 9. Simplify $x^6 x^8$.

Problem 10. Simplify $6x^4 x^4$.

Problem 11. Simplify $x^8 x^9$.

Problem 12. Simplify $10x^5 x^5$.

Problem 13. Simplify $x^3 x^4 x^5$.

Problem 14. Simplify $x^7 x^8 x$.

Problem 15. Simplify $6x^2 x^2 x^2$.

Problem 16. Simplify $4x^5 x x^3$.

Problem 17. Simplify $x^9 x^8 x^6$.

Problem 18. Simplify $9x x^8 x$.

Problem 19. Simplify $5x^4 x^3 x^2 x^2$.

Problem 20. Simplify $x^5 x^7 x^4 x$.

The rule for multiplying variables that have exponents can be extended to include parentheses. For example, $(x^2)^3$ means $x^2x^2x^2 = x^{2+2+2} = x^6$, which is equivalent to $x^{(2)(3)} = x^6$. This is an example of the rule $(x^m)^n = x^{mn}$. In this rule, m multiplies n. (Don't confuse this with the previous rule, $x^mx^n = x^{m+n}$, where m is added to n.)

As another example, $(2x)^4$ means $(2x)(2x)(2x)(2x) = 2^4x^4 = 16x^4$ because $2^4 = (2)(2)(2)(2) = 16$. This is an example of the rule $(ax)^n = a^nx^n$.

The rules $(x^m)^n = x^{mn}$ and $(ax)^n = a^nx^n$ can be combined together to form the rule $(ax^m)^n = a^nx^{mn}$. For example, $(10x^5)^3 = 10^3x^{5(3)} = 1000x^{15}$

Chapter 2 Examples (Set 2)

Example 4. Simplify $(x^3)^4$.
Multiply the exponents: $(x^3)^4 = x^{(3)(4)} = x^{12}$.

Example 5. Simplify $(4x)^3$.
The exponent applies to each base: $(4x)^3 = 4^3x^3 = 64x^3$. Note: $4^3 = (4)(4)(4)$.

Example 6. Simplify $(2x^2)^4$.
Apply the exponent to each factor: $(2x^2)^4 = 2^4x^{(2)(4)} = 16x^8$. Note: $2^4 = (2)(2)(2)(2)$.

Example 7. Simplify $(3x^4)^2(2x^2)^3$.
Apply multiple rules: $(3x^4)^2(2x^2)^3 = 3^2x^{(4)(2)}2^3x^{(2)(3)} = 9x^88x^6 = 72x^{8+6} = 72x^{14}$. Notes: $2^3 = (2)(2)(2) = 8$ and $9x^88x^6$ means 9 times x^8 times 8 times x^6. Since the order in which numbers are multiplied does not matter, this may be written as 9 times 8 times x^8 times x^6, which equals 72 times x^{8+6}.

Chapter 2 Problems (Set 2)

Directions: Apply the rules for multiplying exponents to simplify each expression.

Problem 21. Simplify $(x^5)^2$.

Problem 22. Simplify $(x^2)^4$.

Problem 23. Simplify $(x^3)^3$.

Problem 24. Simplify $(x^7)^5$.

Problem 25. Simplify $(x^4)^8$.

Problem 26. Simplify $(x^9)^6$.

Problem 27. Simplify $(3x)^3$.

Problem 28. Simplify $(6x)^2$.

Problem 29. Simplify $(2x)^5$.

Problem 30. Simplify $(5x)^3$.

Problem 31. Simplify $(7x)^2$.

Problem 32. Simplify $(3x)^4$.

Problem 33. Simplify $(5x^3)^2$.

Problem 34. Simplify $(2x^4)^3$.

Problem 35. Simplify $(10x^5)^4$.

Problem 36. Simplify $(4x^3)^3$.

Problem 37. Simplify $(8x^4)^2$.

Problem 38. Simplify $(2x^9)^4$.

Problem 39. Simplify $(5x)^2(4x^2)^3$.

Problem 40. Simplify $(x^2)^4(6x^5)^2$.

Distribute to Binomials

A **polynomial** is an expression that has terms of the form ax^b added together, where a and b are constants. Examples of polynomials include $x^2 + 6x - 4$ and $3x^8 - 5x^5$. A polynomial may include a constant term, like $2x^7 + 5$, but it does not need to. A polynomial is usually arranged with the terms in order from the greatest exponent to the smallest exponent. For example, in the polynomial $2x^3 - 4x^2 + 9x + 5$, the first term has an exponent of 3, the second term has an exponent of 2, the next term has an exponent of 1 (because $x = x^1$), and the constant corresponds to an exponent of 0 (because of the important rule for exponents that $x^0 = 1$, provided that x is nonzero).

A **binomial** is a polynomial that consists of exactly two terms, like $2x^4 - 5x$ or $x^2 + 4$.

According to the **distributive property**, $a(b + c) = ab + ac$. It is easy to verify that the distributive property for various combinations of numbers. For example, $6(2 + 3) = 6(2) + 6(3)$ because the left-hand side equals $6(5) = 30$ and the right-hand side also equals $6(2) + 6(3) = 12 + 18 = 30$. It also works if there is a minus sign. For example, $3(9 - 2) = 3(9) + 3(-2)$ because $3(9 - 2) = 3(7) = 21$ agrees with $3(9) + 3(-2) = 27 - 6 = 21$.

The distributive property also applies if a, b, and c are algebraic expressions. As an example, $2x(3x^2 - 5x) = 2x(3x^2) + 2x(-5x) = 6x^3 - 10x^2$. Note that it would be perfectly fine to write this as $2x(3x^2 - 5x) = 2x(3x^2) - 2x(5x) = 6x^3 - 10x^2$.

Chapter 3 Examples

Example 1. Distribute $3(2x + 4)$.
Distribute 3 to both terms: $3(2x + 4) = 3(2x) + 3(4) = 6x + 12$.

Example 2. Distribute $4x(5x^2 - 2)$.
Distribute $4x$ to both terms: $4x(5x^2 - 2) = 4x(5x^2) + 4x(-2) = 20x^3 - 8x$.
Alternative solution: $4x(5x^2 - 2) = 4x(5x^2) - 4x(2) = 20x^3 - 8x$.

Example 3. Distribute $2x^3(6x^4 + 2x^3)$.
Distribute $2x^3$ to both terms: $2x^3(6x^4 + 2x^3) = 2x^3(6x^4) + 2x^3(2x^3) = 12x^7 + 4x^6$.

Chapter 3 Problems

Directions: Apply the distributive property.

Problem 1. Distribute $9(x + 6)$.

Problem 2. Distribute $x(x - 4)$.

Problem 3. Distribute $3x(-x^2 + 7x)$.

Problem 4. Distribute $6x^3(2x + 5)$.

Problem 5. Distribute $4x^2(-5x^2 - 3)$.

Problem 6. Distribute $x^4(x^2 - 8x)$.

Problem 7. Distribute $2x(4x^3 + 6x^2)$.

Problem 8. Distribute $x^2(-x^2 + 5)$.

Problem 9. Distribute $4(7x^3 + x)$.

Problem 10. Distribute $3x^2(-6x^4 - 5x^2)$.

Problem 11. Distribute $x(x + 9)$.

Problem 12. Distribute $8x^3(-6x^3 + 7x)$.

Problem 13. Distribute $5x(x - 2)$.

Problem 14. Distribute $2x(-3x^2 + 6x)$.

Problem 15. Distribute $8(3x^3 + 7x^2)$.

Problem 16. Distribute $x(-x^4 - x)$.

Problem 17. Distribute $5x^2(6x^3 + 5)$.

Problem 18. Distribute $2x(-x^2 - 7x)$.

Problem 19. Distribute $7(3x^4 - 6x^2)$.

Problem 20. Distribute $6x^3(-5x^2 + 9x)$.

Problem 21. Distribute $4x^2(8x^2 - 3)$.

Problem 22. Distribute $5x^3(-7x^6 - 4x^2)$.

Problem 23. Distribute $x^2(-x^3 + x)$.

Problem 24. Distribute $9x(5x^2 + 2x)$.

Problem 25. Distribute $7x^2(-3x^3 + 4x^2)$.

Problem 26. Distribute $8x^4(7x^5 - 6x^3)$.

Problem 27. Distribute $6x^7(6x^4 + 3x^2)$.

Problem 28. Distribute $4x^5(-x^5 + x^3)$.

Problem 29. Distribute $9x^8(6x^6 - 9)$.

Problem 30. Distribute $5x^6(-7x^9 - 8x^5)$.

Distribute to Trinomials

A **trinomial** is a polynomial that consists of exactly three terms, like $3x^5 - 2x^3 + x$. The distributive property applies to trinomials the same way as it does to binomials, except that there is one more term. For example, $2x^2(4x^4 + 3x^2 - 5) = 2x^2(4x^4) + 2x^2(3x^2) + 2x^2(-5) = 8x^6 + 6x^4 - 10x^2$.

Chapter 4 Examples

Example 1. Distribute $4(3x^2 + 2x + 6)$.
Distribute 4 to each term:
$$4(3x^2 + 2x + 6) = 4(3x^2) + 4(2x) + 4(6) = 12x^2 + 8x + 24$$

Example 2. Distribute $3x(2x^3 - 4x^2 + 7x)$.
Distribute $3x$ to each term:
$$3x(2x^3 - 4x^2 + 7x) = 3x(2x^3) + 3x(-4x^2) + 3x(7x) = 6x^4 - 12x^3 + 21x^2$$

Example 3. Distribute $2x^2(x^4 - 3x^2 - 2)$.
Distribute $2x^2$ to each term:
$$2x^2(x^4 - 3x^2 - 2) = 2x^2(x^4) + 2x^2(-3x^2) + 2x^2(-2) = 2x^6 - 6x^4 - 4x^2$$

Example 4. Distribute $x^3(-5x^5 + 3x^3 - x)$.
Distribute x^3 to each term:
$$x^3(-5x^5 + 3x^3 - x) = x^3(-5x^5) + x^3(3x^3) + x^3(-x) = -5x^8 + 3x^6 - x^4$$

Chapter 4 Problems

Directions: Apply the distributive property.

Problem 1. Distribute $3(7x^2 + 2x + 4)$.

Problem 2. Distribute $5x(4x^2 + 8x - 9)$.

Problem 3. Distribute $2(5x^3 - 3x^2 + 6x)$.

Problem 4. Distribute $x^2(-2x^2 - 4x + 6)$.

Problem 5. Distribute $4x(3x^4 - 6x^2 - 7)$.

Problem 6. Distribute $8(2x^2 + 3x + 9)$.

Problem 7. Distribute $3x^2(7x^2 + 6x - 4)$.

Problem 8. Distribute $7x(-4x^5 + 5x^3 - 8x)$.

Problem 9. Distribute $9(6x^3 + x^2 - 2x)$.

Problem 10. Distribute $2x^4(-3x^2 + 3x + 9)$.

Problem 11. Distribute $x(x^3 - x^2 + 1)$.

Problem 12. Distribute $5x^2(6x^5 - 4x^4 + 7)$.

Problem 13. Distribute $6(x^4 + 9x^2 + 8)$.

Problem 14. Distribute $4x^3(-6x^6 - 5x^3 - 7x^2)$.

Problem 15. Distribute $8x(4x^5 + x^3 + 3)$.

Problem 16. Distribute $x^5(9x^6 - 2x^2 + 3x)$.

Problem 17. Distribute $6x^2(3x^2 + 4x + 5)$.

Problem 18. Distribute $2x(2x^3 + 8x - 9)$.

Problem 19. Distribute $9x^4(-7x^5 - 6x^3 + 7x)$.

Problem 20. Distribute $7(4x^2 + 9x - 4)$.

Problem 21. Distribute $6x^3(-5x^7 - 3x^5 + 8x^4)$.

Problem 22. Distribute $9x(x^3 + 6x^2 + 3)$.

Problem 23. Distribute $x^3(-8x^5 - 2x^4 - x^2)$.

Problem 24. Distribute $5(3x^6 - 9x^4 + 7x^2)$.

Problem 25. Distribute $8x^2(-4x^4 - x^3 - 5)$.

Problem 26. Distribute $5x^3(5x^2 + 7x - 9)$.

Problem 27. Distribute $6x^5(3x^3 - 8x^2 + 7x)$.

Problem 28. Distribute $2x^4(-4x^8 + 5x^5 - x^2)$.

Problem 29. Distribute $6x(9x^5 - 7x^3 - 5x)$.

Problem 30. Distribute $3x^7(2x^2 + 8x + 4)$.

Problem 31. Distribute $7x^2(-x^4 - 3x^2 + 7)$.

Problem 32. Distribute $5x^8(4x^6 + 9x^4 - 6)$.

Problem 33. Distribute $8x^3(5x^9 - 8x^6 + 2x^3)$.

Problem 34. Distribute $9x^6(3x^8 - 9x^5 - 4x^2)$.

Distribute to Longer Polynomials

The polynomials in this chapter consist of at least four terms. The same method from the previous two chapters applies to these longer polynomials.

Chapter 5 Examples

Example 1. Distribute $2(4x^3 + x^2 - 5x + 6)$.

Distribute 2 to each term:
$$2(4x^3 + x^2 - 5x + 6) = 2(4x^3) + 2(x^2) + 2(-5x) + 2(6)$$
$$= 8x^3 + 2x^2 - 10x + 12$$

Example 2. Distribute $4x(x^4 - 2x^3 + 3x^2 - 5x)$.

Distribute $4x$ to each term:
$$4x(x^4 - 2x^3 + 3x^2 - 5x) = 4x(x^4) + 4x(-2x^3) + 4x(3x^2) + 4x(-5x)$$
$$= 4x^5 - 8x^4 + 12x^3 - 20x^2$$

Example 3. Distribute $3x^2(4x^6 - 3x^4 - 5x^2 + 8)$.

Distribute $3x^2$ to each term:
$$3x^2(4x^6 - 3x^4 - 5x^2 + 8) = 3x^2(4x^6) + 3x^2(-3x^4) + 3x^2(-5x^2) + 3x^2(8)$$
$$= 12x^8 - 9x^6 - 15x^4 + 24x^2$$

Chapter 5 Problems

Directions: Apply the distributive property.

Problem 1. Distribute $5(5x^3 + x^2 + 6x + 4)$.

Problem 2. Distribute $3x(6x^7 - 4x^5 + 3x^3 - 2x)$.

Problem 3. Distribute $x^3(4x^7 + 8x^5 + 6x^3 - 3x)$.

Problem 4. Distribute $2x^2(-9x^4 - 6x^3 + 3x^2 + x)$.

Problem 5. Distribute $8(4x^6 - 6x^4 - 7x^2 + 9)$.

Problem 6. Distribute $9x^3(x^3 + 3x^2 - 6x - 9)$.

Problem 7. Distribute $6x^5(-5x^6 + 3x^3 + 4x^2 + 5x)$.

Problem 8. Distribute $4x(7x^3 + 6x^2 + 5x + 4)$.

Problem 9. Distribute $5x^4(7x^9 - 8x^7 - 6x^5 - 3x^3)$.

Problem 10. Distribute $7x^6(2x^7 - 4x^6 + 6x^4 - 10)$.

Problem 11. Distribute $4x^7(-9x^6 + 6x^5 - 9x^4 + x^3)$.

Problem 12. Distribute $9x^4(-7x^7 - 6x^5 - 5x^4 - 7x^2)$.

Problem 13. Distribute $6x(-8x^9 - 2x^6 - x^5 + 100)$.

Problem 14. Distribute $4x^3(-6x^9 + 4x^8 - 9x^6 - 5x^3 - 7x)$.

Problem 15. Distribute $4x^4(7x^7 - 6x^6 - 5x^5 - 4x^4 + 8x^3)$.

Problem 16 $10x^7(-8x^7 + 4x^6 + 10x^5 - 9x^2 - 12x - 5)$.

Problem 17 $8x^9(2x^{13} + 6x^{11} - 3x^9 - 7x^7 + 9x^5 - 4x^3 + 8x)$.

Distribute Minus Signs

When a minus sign appears with a factor that multiplies parentheses, the minus sign distributes to each term in parentheses. For example, $-a(b + c) = -ab - ac$. If there are any minus signs inside of the parentheses, note that the product of two negative numbers is positive. For example, $-a(b - c) = -a(b) - a(-c) = -ab + ac$. When a minus sign comes before parentheses, pay careful attention to all of the signs.

Chapter 6 Examples

Example 1. Distribute $-4(x + 3)$.
Distribute -4 to both terms: $-4(x + 3) = -4(x) - 4(3) = -4x - 12$.

Example 2. Distribute $-2x(3x^2 - 6)$.
Distribute $-2x$ to both terms: $-2x(3x^2 - 6) = -2x(3x^2) - 2x(-6) = -6x^3 + 12x$.

Example 3. Distribute $-3x^2(-2x^2 + 5x - 4)$.
Distribute $-3x^2$ to each term:
$$-3x^2(-2x^2 + 5x - 4) = -3x^2(-2x^2) - 3x^2(5x) - 3x^2(-4) = 6x^4 - 15x^3 + 12x^2$$
Tip: Each term of the polynomial changes sign.

Example 4. Distribute $-x^3(x^4 + 6x^2 - 3)$.
Distribute $-x^3$ to each term:
$$-x^3(x^4 + 6x^2 - 3) = -x^3(x^4) - x^3(6x^2) - x^3(-3) = -x^7 - 6x^5 + 3x^3$$

Chapter 6 Problems

Directions: Apply the distributive property.

Problem 1. Distribute $-5(9x + 4)$.

Problem 2. Distribute $-x(4x^2 - 2)$.

Problem 3. Distribute $-4x^2(3x^3 + 9x)$.

Problem 4. Distribute $-(-x - 6)$.

Problem 5. Distribute $-6x(-8x + 1)$.

Problem 6. Distribute $-5x^3(7x^2 + 6x)$.

Problem 7. Distribute $-x(x - 2)$.

Problem 8. Distribute $-3x(-3x^3 - 4x^2)$.

Problem 9. Distribute $-x^2(6x^2 + 5x)$.

Problem 10. Distribute $-6(4x^2 - 8x)$.

Problem 11. Distribute $-9x(7x^3 + 8x)$.

Problem 12. Distribute $-2x^3(-4x^3 - 12)$.

Problem 13. Distribute $-6x^2(-5x^4 + 4x^2)$.

Problem 14. Distribute $-8x^4(6x^2 + 5x)$.

Problem 15. Distribute $-7x^2(3x - 5)$.

Problem 16. Distribute $-10x^5(-9x^4 - 10x^3)$.

Problem 17. Distribute $-3(x^2 + 5x + 9)$.

Problem 18. Distribute $-(2x^4 - 7x^2 + 9)$.

Problem 19. Distribute $-8x^2(-x^3 - 7x^2 - 6x)$.

Problem 20. Distribute $-6x(-9x^2 + 8x - 5)$.

Problem 21. Distribute $-5x^3(6x^5 + 2x^3 + 8x)$.

Problem 22. Distribute $-x(-4x^6 - 5x^5 - x)$.

Problem 23. Distribute $-9(-2x^2 - 6x + 7)$.

Problem 24. Distribute $-2x^2(-7x^4 - 8x^3 + 4x^2)$.

Problem 25. Distribute $-4x^5(6x^5 - 6x^4 + 4x)$.

Problem 26. Distribute $-9x^2(-4x^7 - 3x^2 - 5x)$.

Problem 27. Distribute $-6x(2x^3 + x^2 + 4x)$.

Problem 28. Distribute $-7x^2(-3x^3 + 2x - 6)$.

Problem 29. Distribute $-8x(x^3 - 4x^2 + 7x)$.

Problem 30. Distribute $-9x^5(-6x^6 + 5x^3 - 8x)$.

Problem 31. Distribute $-2x^3(-9x^7 + 3x^5 - 5x^3 + x)$.

Problem 32. Distribute $-3x^2(-5x^4 + 6x^3 - 7x^2 - 2x)$.

Problem 33. Distribute $-5x^4(8x^3 + 6x^2 + x - 5)$.

Problem 34. Distribute $-9x^9(-11x^8 + 10x^7 + 15x^5 - 9x^2)$.

Distribute with Fractions

The distributive property also applies when fractions are present. A straightforward example is $\frac{2}{3}(6x + 9) = \frac{2}{3}(6x) + \frac{2}{3}(9) = 4x + 6$. To multiply a fraction and a whole number, the whole number is multiplied by the numerator. The reason is that a whole number can be written as a fraction by dividing by one. For example, $6 = \frac{6}{1}$.

$$\frac{2}{3}(6) = \frac{2(6)}{3} = \frac{12}{3} = 4 \quad \text{and} \quad \frac{2}{3}(9) = \frac{2(9)}{3} = \frac{18}{3} = 6$$

If there are fractions both inside and outside of the parentheses, it will be necessary to multiply fractions. For example, $\frac{3}{4}\left(\frac{5}{6}x^2 - \frac{8}{9}x\right) = \frac{3}{4}\left(\frac{5}{6}x^2\right) + \frac{3}{4}\left(-\frac{8}{9}x\right) = \frac{5}{8}x^2 - \frac{2}{3}x$. To multiply fractions, multiply the numerators together and multiply the denominators together. Divide 15 and 24 each by 3 in order to reduce $\frac{15}{24}$ to $\frac{5}{8}$. Divide 24 and 36 each by 12 to reduce $\frac{24}{36}$ to $\frac{2}{3}$. **Tip**: The multiplication below is a little simpler if you note that $\frac{3}{6} = \frac{1}{2}$ such that $\frac{3(5)}{4(6)} = \frac{1(5)}{4(2)} = \frac{5}{8}$. Similarly, since $\frac{3}{9} = \frac{1}{3}$ and $\frac{8}{4} = \frac{2}{1}$, we get $\frac{3(8)}{4(9)} = \frac{1(2)}{1(3)} = \frac{2}{3}$.

$$\frac{3}{4}\left(\frac{5}{6}\right) = \frac{3(5)}{4(6)} = \frac{15}{24} = \frac{5}{8} \quad \text{and} \quad \frac{3}{4}\left(\frac{8}{9}\right) = \frac{3(8)}{4(9)} = \frac{24}{36} = \frac{2}{3}$$

It is important to note that fractions can be written in a variety of equivalent forms in algebra. For example, all of the expressions below are equivalent. (The expression on the right is NOT a mixed number. It is 2 times the quantity $x + 2$ divided by 5.)

$$\frac{2(x + 2)}{5} = \frac{2}{5}(x + 2) = (x + 2)\frac{2}{5} = 2\frac{x + 2}{5}$$

Chapter 7 Examples

Example 1. Distribute $\frac{4}{3}(9x^2 + 6x)$.

Distribute $\frac{4}{3}$ to both terms: $\frac{4}{3}(9x^2 + 6x) = \frac{4}{3}(9x^2) + \frac{4}{3}(6x) = 12x^2 + 8x$.

Notes: $\frac{4}{3}(9) = \frac{4(9)}{3} = \frac{36}{3} = 12$ and $\frac{4}{3}(6) = \frac{4(6)}{3} = \frac{24}{3} = 8$.

Example 2. Distribute $\frac{3x}{5}\left(\frac{x}{2} - \frac{3}{4}\right)$.

Distribute $\frac{3x}{5}$ to both terms: $\frac{3x}{5}\left(\frac{x}{2} - \frac{3}{4}\right) = \frac{3x}{5}\left(\frac{x}{2}\right) + \frac{3x}{5}\left(-\frac{3}{4}\right) = \frac{3}{10}x^2 - \frac{9}{20}x$.

Notes: $xx = x^1 x^1 = x^2$ (recall Chapter 2), $\frac{3x^2}{10} = \frac{3}{10}x^2$, and $\frac{9x}{20} = \frac{9}{20}x$.

Example 3. Distribute $-\frac{2}{3}\left(\frac{9}{4}x^2 - \frac{3}{2}x\right)$.

Distribute $-\frac{2}{3}$ to both terms: $-\frac{2}{3}\left(\frac{9}{4}x^2 - \frac{3}{2}x\right) = -\frac{2}{3}\left(\frac{9}{4}x^2\right) - \frac{2}{3}\left(-\frac{3}{2}x\right) = -\frac{3}{2}x^2 + x$.

Notes: $\frac{2}{3}\left(\frac{9}{4}\right) = \frac{2(9)}{3(4)} = \frac{18}{12} = \frac{3}{2}$ and $-\frac{2}{3}\left(-\frac{3}{2}\right) = \frac{2(3)}{3(2)} = \frac{6}{6} = 1$.

Example 4. Distribute $\frac{8x^3 + 6x^2}{2}$.

Distribute $\frac{1}{2}$ to both terms: $\frac{8x^3 + 6x^2}{2} = \frac{1}{2}(8x^3 + 6x^2) = \frac{1}{2}(8x^3) + \frac{1}{2}(6x^2) = 4x^3 + 3x^2$.

Example 5. Distribute $\frac{3x(x^2 - 5)}{4}$.

Distribute $\frac{3x}{4}$ to both terms: $\frac{3x(x^2 - 5)}{4} = \frac{3x}{4}(x^2 - 5) = \frac{3x}{4}(x^2) + \frac{3x}{4}(-5) = \frac{3}{4}x^3 - \frac{15}{4}x$.

Example 6. Distribute $4\frac{x^2 - 5x - 6}{3}$.

Distribute $\frac{4}{3}$ to each term: $4\frac{x^2 - 5x - 6}{3} = \frac{4}{3}(x^2 - 5x - 6) = \frac{4}{3}(x^2) + \frac{4}{3}(-5x) + \frac{4}{3}(-6) = \frac{4}{3}x^2 - \frac{20}{3}x - 8$.

Note: $\frac{4}{3}(-6) = \frac{4(-6)}{3} = -\frac{24}{3} = -8$.

Chapter 7 Problems

Directions: Apply the distributive property.

Problem 1. Distribute $\frac{2}{3}(12x + 18)$.

Problem 2. Distribute $\frac{1}{4}(36x^2 - 28x)$.

Problem 3. Distribute $-\frac{3x}{2}(3x + 5)$.

Problem 4. Distribute $\frac{x}{3}(-x^2 + 6x)$.

Problem 5. Distribute $-\frac{4x^2}{5}\left(\frac{10}{3}x - \frac{5}{2}\right)$.

Problem 6. Distribute $\frac{5}{4}\left(-\frac{6}{5}x^3 - \frac{18}{7}x\right)$.

Problem 7. Distribute $\frac{5x}{2}\left(\frac{9}{10}x^2 - \frac{6}{25}x\right)$.

Problem 8. Distribute $-\frac{x^3}{6}(9x^4 - 3x^2 + 15)$.

Problem 9. Distribute $\frac{9x}{4}\left(-\frac{x^2}{3} + \frac{5x}{6} - \frac{10}{27}\right)$.

Problem 10. Distribute $\frac{3(4x-6)}{2}$.

Problem 11. Distribute $-4\frac{9x^2+2}{3}$.

Problem 12. Distribute $\frac{2x(-3x+4)}{5}$.

Problem 13. Distribute $\frac{9x^3-6x}{4}$.

Problem 14. Distribute $\frac{4x^2+2x-8}{2}$.

Problem 15. Distribute $-\frac{x(x-6)}{3}$.

Problem 16. Distribute $x\frac{9x^4+2x^2-12x}{3}$.

Problem 17. Distribute $\frac{3x^2(6x^5-3x^3+x)}{2}$.

Problem 18. Distribute $-9\frac{6x^2-10}{4}$.

Problem 19. Distribute $5x\frac{2x^2-x+3}{9}$.

Divide Variables with Exponents

When x^m is divided by x^n, the exponents get subtracted: $\frac{x^m}{x^n} = x^{m-n}$. For example, if $m = 5$ and $n = 2$, this formula gives $\frac{x^5}{x^2} = x^3$. You can verify that this rule works by trying different numbers. For example, let $x = 10$, for which $10^5 = 100{,}000$, $10^2 = 100$, and $10^3 = 1000$. In this example, $\frac{10^5}{10^2} = 10^3$ because $\frac{100{,}000}{100} = 1000$.

Recall that $x^1 = x$. If no exponent is visible, the exponent equals one.

Note that $x^0 = 1$ (provided that x is nonzero). To understand why $x^0 = 1$, consider the case where $n = m$. In this case, $\frac{x^m}{x^n} = x^{m-n}$ becomes $\frac{x^m}{x^m} = x^{m-m}$. The left-hand side of this equation is 1 because any (nonzero) number divided by itself equals one. The right-hand side of this equation is x^0, which shows that x^0 equals one.

A negative exponent means the following: $x^{-n} = \frac{1}{x^n}$. For example, $2^{-3} = \frac{1}{2^3} = \frac{1}{8}$. One way to understand why $x^{-n} = \frac{1}{x^n}$ is to set $m = 0$ in the formula $\frac{x^m}{x^n} = x^{m-n}$. This gives $\frac{x^0}{x^n} = x^{0-n}$. The left-hand side of this equation is $\frac{1}{x^n}$ because $x^0 = 1$. The right-hand side of this equation is x^{-n}, which shows that x^{-n} equals $\frac{1}{x^n}$.

Chapter 8 Examples

Example 1. Simplify $\frac{x^7}{x^3}$.

Subtract the exponents: $\frac{x^7}{x^3} = x^{7-3} = x^4$.

Example 2. Simplify $\frac{x}{x^5}$.

Recall that $x = x^1$. Subtract the exponents: $\frac{x}{x^5} = \frac{x^1}{x^5} = x^{1-5} = x^{-4} = \frac{1}{x^4}$.

Example 3. Simplify $\frac{x^3 x^2}{x^5}$.

Combine the rules from Chapters 2 and 8: $\frac{x^3 x^2}{x^5} = \frac{x^{3+2}}{x^5} = \frac{x^5}{x^5} = x^{5-5} = x^0 = 1$.

Chapter 8 Problems

Directions: Apply the rules for exponents to simplify each expression.

Problem 1. Simplify $\frac{x^8}{x^2}$.

Problem 2. Simplify $\frac{x^9}{x^4}$.

Problem 3. Simplify $\frac{x^6}{x}$.

Problem 4. Simplify $\frac{x^7}{x^9}$.

Problem 5. Simplify $\frac{x^6}{x^3}$.

Problem 6. Simplify $\frac{x^7}{x^6}$.

Problem 7. Simplify $\frac{x}{x^2}$.

Problem 8. Simplify $\frac{x^3}{x^3}$.

Problem 9. Simplify $\frac{x^7 x^5}{x^4}$.

Problem 10. Simplify $\frac{x^2 x^2}{x^9}$.

Problem 11. Simplify $\frac{x^8 x}{x^5}$.

Problem 12. Simplify $\frac{x^9}{x^6 x^2}$.

Problem 13. Simplify $\frac{x^{12}}{x^4 x^2}$.

Problem 14. Simplify $\frac{x^3 x}{x^4}$.

Problem 15. Simplify $\frac{x^2}{x^9 x}$.

Problem 16. Simplify $\frac{x^7 x^4}{x^8 x^2}$.

Problem 17. Simplify $\frac{x^9 x^8}{x^7 x^6}$.

Problem 18. Simplify $\frac{x^4 x}{x^3 x^3}$.

Problem 19. Simplify $\dfrac{x^6 x^3}{x^8 x^7}$.

Problem 20. Simplify $\dfrac{x^6 x^5 x^4}{x^3}$.

Problem 21. Simplify $\dfrac{x^8}{x^9 x^7 x^5}$.

Problem 22. Simplify $\dfrac{x^{10}}{x^3 x^2 x}$.

Problem 23. Simplify $\dfrac{x^9 x^6 x^3}{x}$.

Problem 24. Simplify $\dfrac{x^8 x^6 x^2}{x^5 x^3}$.

Problem 25. Simplify $\dfrac{x^9 x^8}{x^5 x^3 x}$.

Problem 26. Simplify $\dfrac{x^9 x^9 x^9}{x^7 x^7 x^7}$.

Problem 27. Simplify $\dfrac{x^6 x^4 x^2}{x^7 x^5 x^5}$.

Problem 28. Simplify $\dfrac{x^8 x^4 x}{x^9 x^7 x^6}$.

Factor Binomials

<u>Factoring</u> is basically the distributive property in reverse. When we write $a(b + c) = ab + ac$, we say that the distributive property has been applied. If we write the same thing in reverse, $ab + ac = a(b + c)$, we say that we have factored out the a. When two terms have a common factor, parentheses can be used to pull the common factor out. For example, in the expression $15x^2 + 9x$, each term is evenly divisible by 3 and each term has at least one x, such that $3x$ can be factored out: $15x^2 + 9x = 3x(5x + 3)$.

To factor an expression, follow these steps:
- First find the greatest common factor (GCF) of the coefficients. For example, given $12x^5 + 20x^3$, the coefficients are 12 and 20. The GCF of 12 and 20 equals 4 because 4 is the largest integer that evenly divides into both 12 and 20. Note that $4(3) = 12$ and $4(5) = 20$.
- Now look at the exponent of the variable in each term. Which exponent is the smallest? This power of the variable is common to every term. For example, given $12x^5 + 20x^3$, the first term has x^5 and the second term has x^3. Since 3 is smaller than 5, the expression x^3 can be factored out of both terms.
- Use the answers to the previous steps to determine what can be factored out. For example, the expression $4x^3$ can be factored out of $12x^5 + 20x^3$.
- Divide each term by the expression that will be factored out. For example, $\frac{12x^5}{4x^3} = 3x^2$ and $\frac{20x^3}{4x^3} = 5$. In this example, $3x^2$ and 5 will go in the parentheses.
- Put all of these ideas together to write the factored expression. For example, $12x^5 + 20x^3$ can be factored as $4x^3(3x^2 + 5)$.

You can check the answer using the distributive property. For example, to check that $12x^5 + 20x^3$ can be factored as $4x^3(3x^2 + 5)$, apply the distributive property to the answer and verify that it agrees with the given expression.

$$4x^3(3x^2 + 5) = 4x^3(3x^2) + 4x^3(5) = 12x^5 + 20x^3$$

Recall that a **binomial** is a polynomial that has exactly two terms, like $3x^2 - 2$.

Chapter 9 Examples

Example 1. Factor $6x^4 - 15x^3$.

The GCF of 6 and 15 is equal to 3 (since 3 is the largest integer that evenly divides into both 6 and 15) and x^3 has the smallest exponent. When each term is divided by $3x^3$, we get $\frac{6x^4}{3x^3} - \frac{15x^3}{3x^3} = 2x - 5$. The answer is $6x^4 - 15x^3 = 3x^3(2x - 5)$.

Check the answer: $3x^3(2x - 5) = 3x^3(2x) + 3x^3(-5) = 6x^4 - 15x^3$.

Example 2. Factor $27x^2 + 18$.

The GCF of 27 and 18 is equal to 9 (since 9 is the largest integer that evenly divides into both 27 and 18). Since the term 18 does NOT have a variable, no variable will be factored out. When each term is divided by 9, we get $\frac{27x^2}{9} + \frac{18}{9} = 3x^2 + 2$. The answer is $27x^2 + 18 = 9(3x^2 + 2)$.

Check the answer: $9(3x^2 + 2) = 9(3x^2) + 9(2) = 27x^2 + 18$.

Example 3. Factor $32x^8 - 24x^5$.

The GCF of 32 and 24 is equal to 8 (since 8 is the largest integer that evenly divides into both 32 and 24) and x^5 has the smallest exponent. When each term is divided by $8x^5$, we get $\frac{32x^8}{8x^5} - \frac{24x^5}{8x^5} = 4x^3 - 3$. The answer is $32x^8 - 24x^5 = 8x^5(4x^3 - 3)$.

Check the answer: $8x^5(4x^3 - 3) = 8x^5(4x^3) + 8x^5(-3) = 32x^8 - 24x^5$.

Chapter 9 Problems

Directions: Factor out the greatest common expression.

Problem 1. Factor $4x^5 + 6x^2$.

Problem 2. Factor $16x^2 + 40x$.

Problem 3. Factor $12x - 36$.

Problem 4. Factor $8x^4 + 4x^2$.

Problem 5. Factor $14x^5 - 19x^3$.

Problem 6. Factor $7x^2 + 42x$.

Problem 7. Factor $25x^6 - 15x^4$.

Problem 8. Factor $14x^6 + 21x^2$.

Problem 9. Factor $21x - 28$.

Problem 10. Factor $3x^3 + 27x$.

Problem 11. Factor $18x^9 - 24x^5$.

Problem 12. Factor $45x^4 + 15x^3$.

Problem 13. Factor $16x^6 + 20x$.

Problem 14. Factor $8x^6 + 14x^4$.

Problem 15. Factor $48x^4 - 40$.

Problem 16. Factor $2x^2 + x$.

Problem 17. Factor $54x^8 - 42x^3$.

Problem 18. Factor $18x^7 + 15x$.

Problem 19. Factor $24x^7 - 36x^3$.

Problem 20. Factor $40x^5 + 32x^2$.

Problem 21. Factor $18x^8 - 40x^4$.

Problem 22. Factor $48x^9 - 80x^6$.

Factor Trinomials

Recall that a **trinomial** is a polynomial that has exactly three terms, like $x^2 - 4x + 5$. A trinomial can be factored just like a binomial, except that you need to find the GCF of all three terms. For example, the GCF of 12, 28, and 40 equals 4 because 4 is the largest integer that evenly divides into 12, 28, and 40. This allows $12x^4 - 28x^3 + 40x^2$ to be factored as $4x^2(3x^2 - 7x + 10)$.

Chapter 10 Examples

Example 1. Factor $18x^6 + 12x^5 - 30x^4$.

The GCF of 18, 12, and 30 is equal to 6 and x^4 has the smallest exponent. When each term is divided by $6x^4$, we get $\frac{18x^6}{6x^4} + \frac{12x^5}{6x^4} - \frac{30x^4}{6x^4}$. The answer is $18x^6 + 12x^5 - 30x^4$
$= 6x^4(3x^2 + 2x - 5)$.

Check the answer:

$$6x^4(3x^2 + 2x - 5) = 6x^4(3x^2) + 6x^4(2x) + 6x^4(-5) = 18x^6 + 12x^5 - 30x^4$$

Example 2. Factor $15x^8 - 20x^6 + 10x^5$.

The GCF of 15, 20, and 10 is equal to 5 and x^5 has the smallest exponent. When each term is divided by $5x^5$, we get $\frac{15x^8}{5x^5} - \frac{20x^6}{5x^5} + \frac{10x^5}{5x^5}$. The answer is $15x^8 - 20x^6 + 10x^5$
$= 5x^5(3x^3 - 4x + 2)$.

Check the answer:

$$5x^5(3x^3 - 4x + 2) = 5x^5(3x^3) + 5x^5(-4x) + 5x^5(2) = 15x^8 - 20x^6 + 10x^5$$

Chapter 10 Problems

Directions: Factor out the greatest common expression.

Problem 1. Factor $12x^5 + 18x^3 + 4x$.

Problem 2. Factor $18x^9 + 36x^8 - 24x^7$.

Problem 3. Factor $6x^2 - 9x + 12$.

Problem 4. Factor $7x^6 - 3x^4 - 2x^2$.

Problem 5. Factor $11x^5 + 22x^4 + 33x^3$.

Problem 6. Factor $14x^7 + 28x^4 - 42x$.

Problem 7. Factor $12x^4 - 24x + 8$.

Problem 8. Factor $40x^8 - 80x^5 - 48x^2$.

Problem 9. Factor $35x^{11} + 25x^8 - 10x^5$.

Problem 10. Factor $27x^5 + 81x^4 + 9x^3$.

Problem 11. Factor $48x^7 - 36x^5 + 48x^3$.

Problem 12. Factor $24x^8 + 48x^6 - 72x^4$.

Problem 13. Factor $6x^3 - 8x^2 - 9x$.

Problem 14. Factor $72x^5 - 54x^3 + 108x^2$.

Problem 15. Factor $60x^7 - 105x^6 - 75x^5$.

Problem 16. Factor $180x^4 + 240x^2 + 270$.

Problem 17. Factor $105x^{10} + 63x^7 - 168x^3$.

Problem 18. Factor $16x^7 - 32x^6 + 80x^5$.

Problem 19. Factor $56x^5 - 72x^3 - 32x$.

Problem 20. Factor $180x^{14} + 108x^9 - 288x^4$.

Problem 21. Factor $160x^{15} - 80x^{13} - 140x^9$.

Problem 22. Factor $88x^8 + 22x^7 + 66x^6$.

Problem 23. Factor $54x^4 - 72x^2 + 45$.

Problem 24. Factor $28x^{11} + 42x^8 - 56x^5$.

Problem 25. Factor $125x^4 - 225x^3 - 175x^2$.

Problem 26. Factor $36x^{12} + 12x^{10} - 18x^7$.

Problem 27. Factor $18x^7 + 72x^5 + 144x^3$.

Problem 28. Factor $120x^{10} - 200x^9 + 360x^8$.

Problem 29. Factor $256x^4 + 96x^3 - 160x$.

Problem 30. Factor $100x^{10} - 80x^7 - 30x^4$.

Problem 31. Factor $375x^{10} + 525x^6 + 450x^2$.

Problem 32. Factor $56x^9 + 16x^8 - 56x^7$.

Problem 33. Factor $216x^{13} - 216x^8 + 54x^3$.

Problem 34. Factor $288x^{14} - 384x^{11} - 240x^8$.

Factor Longer Polynomials

The polynomials in this chapter consist of at least four terms. The same method from the previous two chapters applies to these longer polynomials.

Chapter 11 Examples

Example 1. Factor $8x^5 - 12x^4 + 28x^3 - 16x^2$.

The GCF of 8, 12, 28, and 16 is equal to 4 and x^2 has the smallest exponent. When each term is divided by $4x^2$, we get $\frac{8x^5}{4x^2} - \frac{12x^4}{4x^2} + \frac{28x^3}{4x^2} - \frac{16^2}{4x^2}$. The answer is $8x^5 - 12x^4 + 28x^3 - 16x^2 = 4x^2(2x^3 - 3x^2 + 7x - 4)$.

Check the answer:
$$4x^2(2x^3 - 3x^2 + 7x - 4) = 4x^2(2x^3) + 4x^2(-3x^2) + 4x^2(7x) + 4x^2(-4)$$
$$= 8x^5 - 12x^4 + 28x^3 - 16x^2$$

Example 2. Factor $36x^9 + 24x^7 - 48x^5 - 72x^3$.

The GCF of 36, 24, 48, and 72 is equal to 12 and x^3 has the smallest exponent. When each term is divided by $12x^3$, we get $\frac{36x^9}{12x^3} + \frac{24x^7}{12x^3} - \frac{48x^5}{12x^3} - \frac{72x^3}{12x^3}$. The answer is $36x^9 + 24x^7 - 48x^5 - 72x^3 = 12x^3(3x^6 + 2x^4 - 4x^2 - 6)$.

Check the answer:
$$12x^3(3x^6 + 2x^4 - 4x^2 - 6) = 12x^3(3x^6) + 12x^3(2x^4) + 12x^3(-4x^2) + 12x^3(-6)$$
$$= 36x^9 + 24x^7 - 48x^5 - 72x^3$$

Chapter 11 Problems

Directions: Factor out the greatest common expression.

Problem 1. Factor $24x^5 + 18x^4 + 6x^3 + 3x^2$.

Problem 2. Factor $24x^7 + 42x^5 - 30x^3 + 54x$.

Problem 3. Factor $30x^3 - 5x^2 + 40x - 25$.

Problem 4. Factor $50x^8 - 40x^7 - 34x^6 - 6x^5$.

Problem 5. Factor $x^7 + x^4 + x^2 + x$.

Problem 6. Factor $24x^{12} + 72x^9 + 36x^6 - 24x^3$.

Problem 7. Factor $96x^8 - 80x^6 + 40x^4 + 64x^2$.

Problem 8. Factor $28x^4 - 36x^3 + 12x^2 - 24x$.

Problem 9. Factor $45x^{10} - 72x^8 - 18x^6 + 36x^4$.

Problem 10. Factor $48x^5 + 96x^4 - 16x^3 + 80x^2$.

Problem 11. Factor $50x^{14} - 150x^{11} - 75x^8 - 175x^5$.

Problem 12. Factor $49x^9 - 70x^7 + 35x^5 + 56x^3$.

Problem 13. Factor $80x^{10} + 100x^9 - 60x^8 - 120x^7$.

Problem 14. Factor $112x^{22} - 28x^{18} - 70x^{14} + 84x^{10}$.

Problem 15. Factor $72x^{11} + 120x^{10} - 192x^9 - 48x^8$.

Problem 16. Factor $192x^7 - 256x^5 - 256x^3 - 96x$.

Problem 17. Factor $108x^{24} + 243x^{19} - 162x^{14} - 189x^9$.

Factor Minus Signs

When a minus sign is factored out of a polynomial, each term of the polynomial changes sign. For example, $-6x^2 + 9x - 3 = -3(2x^2 - 3x + 1)$. Compare $-6x^2 + 9x - 3$ to $2x^2 - 3x + 1$ to see that each term changed sign. Note that $-3(-3) = 9$ because the product of two negative numbers is positive, whereas $-3(2) = -6$ and $-3(1) = -3$ because the product of a negative number and a positive number is negative.

Chapter 12 Examples

Example 1. Factor $-5x^4 - 10x^3 + 20x^2$. Include a minus sign.

The GCF of 5, 10, and 20 is equal to 5 and x^2 has the smallest exponent. When each term is divided by $-5x^2$, we get $-\frac{5x^4}{-5x^2} - \frac{10x^3}{-5x^2} + \frac{20x^2}{-5x^2}$. The answer is $-5x^4 - 10x^3 + 20x^2 = -5x^2(x^2 + 2x - 4)$. Observe that each term changed sign.

Check the answer:
$$-5x^2(x^2 + 2x - 4) = -5x^2(x^2) - 5x^2(2x) - 5x^2(-4) = -5x^4 - 10x^3 + 20x^2$$

Example 2. Factor $6x^7 - 15x^5 - 9x^3$. Include a minus sign.

The GCF of 6, 15, and 9 is equal to 3 and x^3 has the smallest exponent. When each term is divided by $-3x^3$, we get $\frac{6x^7}{-3x^3} - \frac{15x^5}{-3x^3} - \frac{9x^3}{-3x^3}$. The answer is $6x^7 - 15x^5 - 9x^3 = -3x^3(-2x^4 + 5x^2 + 3)$. Observe that each term changed sign.

Check the answer:
$$-3x^3(-2x^4 + 5x^2 + 3) = -3x^3(-2x^4) - 3x^3(5x^2) - 3x^3(3) = 6x^7 - 15x^5 - 9x^3$$

Chapter 12 Problems

Directions: Factor out the greatest common expression <u>**along with a minus sign**</u>.

Problem 1. Factor $-8x^3 - 12x^2$.

Problem 2. Factor $-2x + 8$.

Problem 3. Factor $-30x^5 - 18x^3$.

Problem 4. Factor $-x^8 - x^4$.

Problem 5. Factor $-6x^9 + 21x^6$.

Problem 6. Factor $-36x^2 + 27x$.

Problem 7. Factor $-9x^6 - 5x^5$.

Problem 8. Factor $-15x^5 + 40x^3$.

Problem 9. Factor $-49x^9 - 63x^3$.

Problem 10. Factor $-48x^4 - 40$.

Problem 11. Factor $-36x^8 - 81x^5 - 36x^2$.

Problem 12. Factor $-24x^2 + 48x + 32$.

Problem 13. Factor $-5x^9 + 25x^7 - 15x^5$.

Problem 14. Factor $x^{10} - 3x^7 - 2x^4$.

Problem 15. Factor $-30x^7 - 46x^5 - 22x^3$.

Problem 16. Factor $-64x^4 - 32x^3 + 48x$.

Problem 17. Factor $-49x^8 + 35x^6 - 56x^4$.

Problem 18. Factor $-18x^4 + 54x^3 + 12x^2$.

Problem 19. Factor $4x^8 - 6x^4 - 9$.

Problem 20. Factor $-72x^5 + 24x^3 - 60x$.

Problem 21. Factor $-144x^5 - 162x^4 + 72x^3$.

Problem 22. Factor $-105x^{15} - 63x^{10} - 147x^5$.

Problem 23. Factor $-8x^6 + 16x^5 - 12x^4 + 18x^3$.

Problem 24. Factor $-6x^{12} + 7x^{10} + 2x^8 + x^6$.

Problem 25. Factor $-35x^3 - 21x^2 - 7x - 28$.

Problem 26. Factor $36x^{12} - 24x^8 - 20x^6 - 8x^4$.

Problem 27. Factor $-48x^{10} + 54x^7 + 42x^4 - 30x$.

Problem 28. Factor $-27x^5 + 36x^4 - 81x^3 + 72x^2$.

Problem 29. Factor $35x^7 - 10x^5 - 20x^4 - 35x$.

Problem 30. Factor $-6x^{11} + 24x^9 + 18x^7 + 9x^5$.

Problem 31. Factor $-8x^5 + 56x^4 - 64x^3 + 48x^2$.

Problem 32. Factor $88x^{16} - 55x^{13} - 33x^{10} - 77x^7$.

Problem 33. Factor $-126x^5 + 56x^4 - 70x^3 - 56x^2$.

Problem 34. Factor $57x^{16} - 152x^{12} - 114x^8 - 171x^4$.

The FOIL Method

The FOIL method is one way to multiply an expression of the form $(a + b)(c + d)$. The word FOIL is an abbreviation for "First Outside Inside Last." Multiply the **<u>first</u>** terms to get a times c, multiply the **<u>outside</u>** terms to get a times d, multiply the **<u>inside</u>** terms to get b times c, and multiply the **<u>last</u>** terms to get b times d. Putting these terms together, we get the following formula:

$$(a + b)(c + d) = ac + ad + bc + bd$$

Chapter 13 Examples

Example 1. Expand $(x + 5)(x - 3)$.

Apply the FOIL method:

$$(x + 5)(x - 3) = x(x) + x(-3) + 5(x) + 5(-3)$$
$$= x^2 - 3x + 5x - 15 = x^2 + 2x - 15$$

Note: In the last step, we combined like terms (Chapter 1): $-3x + 5x = 2x$.

Example 2. Expand $(3x - 4)(2x - 7)$.

Apply the FOIL method:

$$(3x - 4)(2x - 7) = 3x(2x) + 3x(-7) - 4(2x) - 4(-7)$$
$$= 6x^2 - 21x - 8x + 28 = 6x^2 - 29x + 28$$

Note: In the last step, we combined like terms (Chapter 1): $-21x - 8x = -29x$.

Chapter 13 Problems

Directions: Apply the FOIL method to expand each expression. Simplify each answer.

Problem 1. Expand $(x + 3)(x + 2)$.

Problem 2. Expand $(x + 6)(x + 4)$.

Problem 3. Expand $(x - 8)(x - 1)$.

Problem 4. Expand $(x + 3)(x - 9)$.

Problem 5. Expand $(x + 9)(x - 3)$.

Problem 6. Expand $(x + 4)(x + 7)$.

Problem 7. Expand $(x - 9)(x + 6)$.

Problem 8. Expand $(x + 7)(x - 5)$.

Problem 9. Expand $(x + 8)(x + 8)$.

Problem 10. Expand $(x - 5)(x - 3)$.

Problem 11. Expand $(x - 4)(x + 8)$.

Problem 12. Expand $(-x + 8)(x + 5)$.

Problem 13. Expand $(6 - x)(8 - x)$.

Problem 14. Expand $(x + 1)(x + 9)$.

Problem 15. Expand $(x - 4)(x + 4)$.

Problem 16. Expand $(-x + 4)(x + 2)$.

Problem 17. Expand $(x - 5)(-x - 5)$.

Problem 18. Expand $(x + 6)(x + 3)$.

Problem 19. Expand $(9 - x)(2 - x)$.

Problem 20. Expand $(-x - 9)(-x - 9)$.

Problem 21. Expand $(x - 3)(x + 4)$.

Problem 22. Expand $(x + 6)(x - 3)$.

Problem 23. Expand $(x - 5)(x - 2)$.

Problem 24. Expand $(x + 9)(x + 7)$.

Problem 25. Expand $(-x + 7)(x + 6)$.

Problem 26. Expand $(-x - 3)(x + 9)$.

Problem 27. Expand $(x + 7)(x - 9)$.

Problem 28. Expand $(x + 6)(x + 5)$.

Problem 29. Expand $(-x + 5)(x + 8)$.

Problem 30. Expand $(-x - 1)(-x - 1)$.

Problem 31. Expand $(x - 5)(-x - 9)$.

Problem 32. Expand $(x + 4)(x + 2)$.

Problem 33. Expand $(2 - x)(7 - x)$.

Problem 34. Expand $(-x - 8)(-x - 9)$.

Problem 35. Expand $(6x + 5)(3x + 4)$.

Problem 36. Expand $(2x + 3)(7x + 9)$.

Problem 37. Expand $(7x + 4)(5x - 9)$.

Problem 38. Expand $(3x - 5)(3x - 6)$.

Problem 39. Expand $(4x - 7)(6x + 8)$.

Problem 40. Expand $(5x + 2)(6x + 4)$.

Problem 41. Expand $(6x + 7)(2x - 9)$.

Problem 42. Expand $(8x + 9)(8x + 9)$.

Problem 43. Expand $(2x - 6)(6x - 2)$.

Problem 44. Expand $(9x + 5)(3x - 7)$.

Problem 45. Expand $(7x + 6)(8x + 9)$.

Problem 46. Expand $(-5x + 3)(4x - 8)$.

Problem 47. Expand $(4 + 7x)(x - 8)$.

Problem 48. Expand $(6x + 3)(6 - 9x)$.

Problem 49. Expand $(8 + 2x)(5x + 7)$.

Problem 50. Expand $(3x + 5)(3x - 5)$.

Problem 51. Expand $(8 + 4x)(7 - 4x)$.

Problem 52. Expand $(-9x - 8)(-6x + 1)$.

Problem 53. Expand $(4 + 5x)(2x - 3)$.

Problem 54. Expand $(8x - 3)(5 - 4x)$.

Problem 55. Expand $(-2x + 9)(3x - 7)$.

Problem 56. Expand $(6 + 7x)(5 - 2x)$.

Problem 57. Expand $(x^2 + x)(x + 1)$.

Problem 58. Expand $(4x^2 - 5)(3x + 6)$.

Problem 59. Expand $(9x^2 - 8)(6x + 7)$.

Problem 60. Expand $(3x^2 + 5)(4x^2 - 6)$.

Problem 61. Expand $(7x^2 - 6x)(4x - 2)$.

Problem 62. Expand $(2x^2 + 2x)(2x + 2)$.

Problem 63. Expand $(7x^2 + 9)(4x^3 - 6)$.

Problem 64. Expand $(x^2 + 1)(x^2 - 1)$.

Problem 65. Expand $(6x + 5)(7x^2 + 3)$.

Problem 66. Expand $(2x^2 - 6x)(2x - 3)$.

Problem 67. Expand $(6x^2 + 7)(8x^2 - 9)$.

Problem 68. Expand $(4x^2 - 2)(3x + 5)$.

Problem 69. Expand $(5x^3 + 3x)(4x^2 - 3x)$.

Problem 70. Expand $(4x^2 + 4x)(6x^2 + 6)$.

Problem 71. Expand $(5x^3 - 9x)(3x - 5)$.

Problem 72. Expand $(7x^2 + 5x)(6x^3 - 5x)$.

Problem 73. Expand $(4x^3 - 7x)(6x^2 + 2)$.

Problem 74. Expand $(2x^2 + 4)(6x^3 - 5)$.

Problem 75. Expand $(9x^3 - 5x)(8x^4 - 4x^2)$.

Problem 76. Expand $(6x^3 + 3x)(4x^2 + 3x)$.

Problem 77. Expand $(-6x + 7)(-5x^3 - 8)$.

Problem 78. Expand $(5x^2 + 3)(4x^3 - 5)$.

Problem 79. Expand $(2x^3 - 6x)(3x^4 + 8x^2)$.

Problem 80. Expand $(4x^3 + 3x)(4x^4 - 3x^2)$.

Problem 81. Expand $(9x^4 - 7x^2)(8x^3 - 6x)$.

Problem 82. Expand $(6x^3 - 9x^2)(7 + 9x^2)$.

Square of the Sum

One way to square an expression of the form $a + b$ is to apply the FOIL method from Chapter 13:
$$(a + b)^2 = (a + b)(a + b) = a(a) + a(b) + b(a) + b(b) = a^2 + 2ab + b^2$$
The above algebra is so common that it is worth remembering the formula (or better yet, remembering the method). The main idea is that each term gets squared (a^2 and b^2) and there is also a very important cross term ($2ab$). Remember the term $2ab$ when expanding an expression that has the form $(a + b)^2$.

Chapter 14 Examples (Set 1)

Example 1. Expand $(x + 4)^2$.
Apply the formula $(a + b)^2 = a^2 + 2ab + b^2$ with $a = x$ and $b = 4$.
$$(x + 4)^2 = x^2 + 2(x)(4) + 4^2 = x^2 + 8x + 16$$

Example 2. Expand $(5x - 3)^2$.
Apply the formula $(a + b)^2 = a^2 + 2ab + b^2$ with $a = 5x$ and $b = -3$.
$$(5x - 3)^2 = (5x)^2 + 2(5x)(-3) + (-3)^2 = 25x^2 - 30x + 9$$
Note: $(5x)^2 = 5^2 x^2 = 25x^2$ (recall the second part of Chapter 2).

Example 3. Expand $(-2x + 1)^2$.
Apply the formula $(a + b)^2 = a^2 + 2ab + b^2$ with $a = -2x$ and $b = 1$.
$$(-2x + 1)^2 = (-2x)^2 + 2(-2x)(1) + 1^2 = 4x^2 - 4x + 1$$

Chapter 14 Problems (Set 1)

Directions: Expand each expression. When possible, simplify the answer.

Problem 1. Expand $(x + 1)^2$.

Problem 2. Expand $(x + 6)^2$.

Problem 3. Expand $(x - 2)^2$.

Problem 4. Expand $(x - 3)^2$.

Problem 5. Expand $(-x + 7)^2$.

Problem 6. Expand $(x + 9)^2$.

Problem 7. Expand $(8 - x)^2$.

Problem 8. Expand $(x - 12)^2$.

Problem 9. Expand $(20 + x)^2$.

Problem 10. Expand $(-x - 5)^2$.

Problem 11. Expand $(3x + 6)^2$.

Problem 12. Expand $(6x - 4)^2$.

Problem 13. Expand $(7x - 6)^2$.

Problem 14. Expand $(4x + 5)^2$.

Problem 15. Expand $(8x - 5)^2$.

Problem 16. Expand $(2x + 7)^2$.

Problem 17. Expand $(9x - 9)^2$.

Problem 18. Expand $(-5x + 8)^2$.

Problem 19. Expand $(7 + 3x)^2$.

Problem 20. Expand $(-6x - 4)^2$.

Problem 21. Expand $(1 - 8x)^2$.

Problem 22. Expand $(-9x + 2)^2$.

Problem 23. Expand $(3x^2 + 4)^2$.

Problem 24. Expand $(8x^2 - 5x)^2$.

Problem 25. Expand $(-6x^3 + 1)^2$.

Problem 26. Expand $(9x^4 + 7x^2)^2$.

Problem 27. Expand $(5x^2 - 5)^2$.

Problem 28. Expand $(-2x^3 - 8x^2)^2$.

Problem 29. Expand $(4 + 3x^2)^2$.

Problem 30. Expand $(7x^3 - 6x)^2$.

Problem 31. Expand $(1 - x^5)^2$.

Problem 32. Expand $(-8x^4 + 4x^2)^2$.

Problem 33. Expand $(x - x^3)^2$.

Problem 34. Expand $(9x^5 - 8x^4)^2$.

An expression that has the form $a^2 + 2ab + b^2$ can be factored by applying the square of the sum formula in reverse:

$$a^2 + 2ab + b^2 = (a + b)^2$$

For example, $16x^2 + 24x + 9$ can be factored by identifying $a = 4x$ and $b = 3$:

$$16x^2 + 24x + 9 = (4x)^2 + 2(4x)(3) + 3^2 = (4x + 3)^2$$

Use the square of the sum formula to check the answer:

$$(4x + 3)^2 = (4x)^2 + 2(4x)(3) + 3^2 = 16x^2 + 24x + 9$$

Chapter 14 Examples (Set 2)

Example 4. Factor $x^2 - 6x + 9$.

Identify $a = x$ and $b = -3$. (One is negative because the cross term, $-6x$, is negative.)

$$x^2 - 6x + 9 = x^2 + 2(x)(-3) + (-3)^2 = (x - 3)^2$$

Use the square of the sum formula to check the answer:

$$(x - 3)^2 = x^2 + 2(x)(-3) + (-3)^2 = x^2 - 6x + 9$$

Example 5. Factor $4x^2 + 20x + 25$.

Identify $a = 2x$ and $b = 5$:

$$4x^2 + 20x + 25 = (2x)^2 + 2(2x)(5) + 5^2 = (2x + 5)^2$$

Use the square of the sum formula to check the answer:

$$(2x + 5)^2 = (2x)^2 + 2(2x)(5) + 5^2 = 4x^2 + 20x + 25$$

Example 6. Factor $9 - 36x + 36x^2$.

Identify $a = 3$ and $b = -6x$. (One is negative because the cross term, $-36x$, is negative.)

$$9 - 36x + 36x^2 = 3^2 + 2(3)(-6x) + (-6x)^2 = (3 - 6x)^2$$

Alternate answers: $(-3 + 6x)^2$ and $(6x - 3)^2$ are equivalent to $(3 - 6x)^2$.

Use the square of the sum formula to check the answer:

$$(3 - 6x)^2 = 3^2 + 2(3)(-6x) + (-6x)^2 = 9 - 36x + 36x^2$$

Chapter 14 Problems (Set 2)

Directions: Factor each expression.

Problem 35. Factor $x^2 + 12x + 36$.

Problem 36. Factor $x^2 - 8x + 16$.

Problem 37. Factor $x^2 + 18x + 81$.

Problem 38. Factor $9x^2 + 24x + 16$.

Problem 39. Factor $16x^2 - 72x + 81$.

Problem 40. Factor $64x^2 - 80x + 25$.

Problem 41. Factor $25x^2 + 70x + 49$.

Problem 42. Factor $49x^2 + 70x + 25$.

Problem 43. Factor $x^4 - 2x^2 + 1$.

Problem 44. Factor $9x^4 + 36x^3 + 36x^2$.

Difference of Squares

When the FOIL method from Chapter 13 is applied to an expression that has the form $(a + b)(a - b)$, the cross term cancels out:

$$(a + b)(a - b) = a(a) + a(-b) + b(a) + b(-b) = a^2 - ab + ab - b^2 = a^2 - b^2$$

The above algebra is so common that it is worth remembering the formula.

Chapter 15 Examples (Set 1)

Example 1. Expand $(x + 3)(x - 3)$.

Apply the formula $(a + b)(a - b) = a^2 - b^2$ with $a = x$ and $b = 3$.

$$(x + 3)(x - 3) = x^2 - 3^2 = x^2 - 9$$

Example 2. Expand $(2x + 4)(2x - 4)$.

Apply the formula $(a + b)(a - b) = a^2 - b^2$ with $a = 2x$ and $b = 4$.

$$(2x + 4)(2x - 4) = (2x)^2 - 4^2 = 4x^2 - 16$$

Note: $(2x)^2 = 2^2 x^2 = 4x^2$ (recall the second part of Chapter 2).

Example 3. Expand $(-3x + 2)(3x + 2)$.

Apply the formula $(a + b)(a - b) = a^2 - b^2$. First note that $-3x + 2 = 2 - 3x$ and $3x + 2 = 2 + 3x$ such that $(-3x + 2)(3x + 2) = (2 - 3x)(2 + 3x)$. Now note that $(2 - 3x)(2 + 3x) = (2 + 3x)(2 - 3x)$. Let $a = 2$ and $b = 3x$.

$$(-3x + 2)(3x + 2) = (2 - 3x)(2 + 3x) = 2^2 - (3x)^2 = 4 - 9x^2 = -9x^2 + 4$$

Chapter 15 Problems (Set 1)

Directions: Expand each expression.

Problem 1. Expand $(x + 2)(x - 2)$.

Problem 2. Expand $(x + 5)(x - 5)$.

Problem 3. Expand $(x + 4)(x - 4)$.

Problem 4. Expand $(4 + x)(4 - x)$.

Problem 5. Expand $(x - 7)(x + 7)$.

Problem 6. Expand $(6 + x)(6 - x)$.

Problem 7. Expand $(x - 5)(x + 5)$.

Problem 8. Expand $(9 + x)(9 - x)$.

Problem 9. Expand $(-x + 1)(-x - 1)$.

Problem 10. Expand $(8 - x)(8 + x)$.

Problem 11. Expand $(5x + 2)(5x - 2)$.

Problem 12. Expand $(9x + 7)(9x - 7)$.

Problem 13. Expand $(3x - 4)(3x + 4)$.

Problem 14. Expand $(6x + 3)(6x - 3)$.

Problem 15. Expand $(8x - 6)(8x + 6)$.

Problem 16. Expand $(4x + 4)(4x - 4)$.

Problem 17. Expand $(7 + 2x)(7 - 2x)$.

Problem 18. Expand $(9x + 5)(9x - 5)$.

Problem 19. Expand $(-6x + 3)(6x + 3)$.

Problem 20. Expand $(5 - 4x)(5 + 4x)$.

Problem 21. Expand $(-8x - 9)(-8x + 9)$.

Problem 22. Expand $(7x + 6)(7x - 6)$.

Problem 23. Expand $(x^2 + 1)(x^2 - 1)$.

Problem 24. Expand $(x^2 + 7)(x^2 - 7)$.

Problem 25. Expand $(x^3 - 2)(x^3 + 2)$.

Problem 26. Expand $(x^3 - 4)(x^3 + 4)$.

Problem 27. Expand $(5x^3 + 3)(5x^3 - 3)$.

Problem 28. Expand $(10x^4 + 10)(10x^4 - 10)$.

Problem 29. Expand $(7x^4 - 7)(7x^4 + 7)$.

Problem 30. Expand $(2x^5 + 5x^3)(2x^5 - 5x^3)$.

Problem 31. Expand $(4x^6 + 6x^4)(4x^6 - 6x^4)$.

Problem 32. Expand $(9 - 6x^6)(9 + 6x^6)$.

Problem 33. Expand $(6x^7 + 2)(6x^7 - 2)$.

Problem 34. Expand $(-10 + 8x^8)(-10 - 8x^8)$.

An expression that has the form $a^2 - b^2$ can be factored by applying the difference of squares formula in reverse:

$$a^2 - b^2 = (a + b)(a - b)$$

For example, $4x^2 - 9$ can be factored by identifying $a = 2x$ and $b = 3$:

$$4x^2 - 9 = (2x)^2 - 3^2 = (2x + 3)(2x - 3)$$

Use the difference of squares formula to check the answer:

$$(2x + 3)(2x - 3) = (2x)^2 - 3^2 = 4x^2 - 9$$

Chapter 14 Examples (Set 2)

Example 4. Factor $x^2 - 16$.

Identify $a = x$ and $b = 4$.

$$x^2 - 16 = x^2 - 4^2 = (x + 4)(x - 4)$$

Use the difference of squares formula to check the answer:

$$(x + 4)(x - 4) = x^2 - 4^2 = x^2 - 16$$

Example 5. Factor $9x^2 - 25$.

Identify $a = 3x$ and $b = 5$:

$$9x^2 - 25 = (3x)^2 - 5^2 = (3x + 5)(3x - 5)$$

Use the difference of squares formula to check the answer:

$$(3x + 5)(3x - 5) = (3x)^2 - 5^2 = 9x^2 - 25$$

Example 6. Factor $16 - 4x^6$.

Identify $a = 4$ and $b = 2x^3$. The idea is that $(2x^3)^2 = 2^2 x^6 = 4x^6$ (recall Chapter 2).

$$16 - 4x^6 = 4^2 - (2x^3)^2 = (4 + 2x^3)(4 - 2x^3)$$

Use the difference of squares formula to check the answer:

$$(4 + 2x^3)(4 - 2x^3) = 4^2 - (2x^3)^2 = 16 - 4x^6$$

Chapter 14 Problems (Set 2)

Directions: Factor each expression.

Problem 35. Factor $x^2 - 4$.

Problem 36. Factor $x^2 - 25$.

Problem 37. Factor $25 - x^2$.

Problem 38. Factor $64 - x^2$.

Problem 39. Factor $x^2 - 36$.

Problem 40. Factor $49x^2 - 81$.

Problem 41. Factor $16x^2 - 121$.

Problem 42. Factor $144 - 81x^2$.

Problem 43. Factor $x^4 - 1$.

Problem 44. Factor $9x^2 - 64$.

Isolate the Unknown

A large class of algebra problems can be solved by <u>isolating the unknown</u> as follows:

- First, bring all of the variable terms to one side of the equation and all of the constant terms to the opposite side of the equation. A term can be moved to the other side of the equation by changing its sign. For example, to bring $2x$ to the left side in $5x = 2x + 12$, subtract $2x$ from both sides to get $5x - 2x = 12$. Observe that $2x$ is now negative after moving to the left side of the equation. As another example, to bring $3x$ to the right side in $28 - 3x = 4x$, add $3x$ to both sides to get $28 = 4x + 3x$. Observe that $3x$ is now positive after moving to the right side of the equation.

- <u>**Combine like terms**</u>. It may help to review Chapter 1. For example, $8x - 5x = 9 + 6$ becomes $3x = 15$ after like terms are combined.

- Once the variable term is by itself, divide by its <u>**coefficient**</u>. For example, divide by 3 on both sides of $3x = 15$ to get $x = \frac{15}{3} = 5$.

- Check the answer by plugging it into the original equation. For example, plug $x = 4$ into $3x + 5 = 6x - 7$. If both sides of the equation are equal, the answer $x = 4$ solves the equation. When x is replaced with 4 in $3x + 5 = 6x - 7$, the left-hand side is $3(4) + 5 = 12 + 5 = 17$ and the right-hand side is $6(4) - 7 = 24 - 7 = 17$. Since both sides equal the same value (17), the answer $x = 4$ is correct.

Be sure to perform the same operation to both sides of the equation. For example, if you divide by 3 on one side of the equation, you must also divide by 3 on the other side.

Chapter 16 Examples

Example 1. Solve the equation $3x = 10 - 2x$.

Bring all of the variable terms to the left side. When $2x$ moves to the left-hand side, it changes sign. (An equivalent way to describe this is: "Add $2x$ to both sides of the equation.")

$$3x + 2x = 10$$

On the left-hand side, combine like terms (Chapter 1): $3x + 2x = 5x$.

$$5x = 10$$

Now that there is only one variable term, divide both sides of the equation by its coefficient.

$$\frac{5x}{5} = \frac{10}{5}$$

On the left-hand side, 5 cancels out because $\frac{5}{5} = 1$. On the right-hand side, 10 divided by 5 equals 2.

$$x = 2$$

Short solution: The solution to this example may be written concisely as follows.

$$3x = 10 - 2x$$
$$5x = 10$$
$$x = \frac{10}{5} = 2$$

Check the answer: Plug $x = 2$ into each side of the original equation. Verify that both sides of the equation equal the same value.

- The left-hand side is $3x = 3(2) = 6$.
- The right-hand side is $10 - 2x = 10 - 2(2) = 10 - 4 = 6$.

Since both sides equal 6, this shows that the answer $x = 2$ correctly solves the given equation.

Example 2. Solve the equation $5 - 3x = 17$.

Bring all of the constant terms to the right side. When 5 moves to the right-hand side, it changes sign. (An equivalent way to describe this is: "Subtract 5 from both sides of the equation.")

$$-3x = 17 - 5$$

Note that the minus sign before $3x$ remains. On the right-hand side, $17 - 5 = 12$.

$$-3x = 12$$

Now that there is only one variable term, divide both sides of the equation by its coefficient. Note that the coefficient is a negative number.

$$\frac{-3x}{-3} = \frac{12}{-3}$$

On the left-hand side, -3 cancels out because $\frac{-3}{-3} = 1$. On the right-hand side, 12 divided by -3 equals -4.

$$x = -4$$

Note: **The student should be familiar with adding, subtracting, multiplying, and dividing negative numbers before learning algebra. Students who are not fluent with negative number arithmetic should review and practice these skills before proceeding.**

Short solution: The solution to this example may be written concisely as follows.

$$5 - 3x = 17$$
$$-3x = 12$$
$$x = \frac{12}{-3} = -4$$

Check the answer: Plug $x = -4$ into each side of the original equation. Verify that both sides of the equation equal the same value.

- The left-hand side is $5 - 3x = 5 - 3(-4) = 5 + 12 = 17$.
- The right-hand side is 17.

Since both sides equal 17, this shows that the answer $x = -4$ correctly solves the given equation.

Example 3. Solve the equation $2x + 5 = 6x - 15$.

Decide whether to put variable terms on the left (and constant terms on the right), or to put variable terms on the right (and constants on the left). In this example, the coefficient of the variable will be positive (after combining like terms) if all of the variable terms are on the right. Bring $2x$ to the right and bring 15 to the left. Each of these terms changes sign when it is brought to the other side. (An equivalent way to describe this is: "Subtract $2x$ from both sides of the equation and add 15 to both sides of the equation.")

$$5 + 15 = 6x - 2x$$

Combine like terms: $5 + 15 = 20$ and $6x - 2x = 4x$.

$$20 = 4x$$

Now that there is only one variable term, divide both sides of the equation by its coefficient.

$$\frac{20}{4} = \frac{4x}{4}$$

On the right-hand side, 4 cancels out because $\frac{4}{4} = 1$. On the left-hand side, 20 divided by 4 equals 5.

$$5 = x$$

The answer may be expressed as $x = 5$. Note that $x = 5$ is equivalent to $5 = x$.

Short solution: The solution to this example may be written concisely as follows.

$$2x + 5 = 6x - 15$$
$$20 = 4x$$
$$\frac{20}{4} = 5 = x$$

Check the answer: Plug $x = 5$ into each side of the original equation. Verify that both sides of the equation equal the same value.

- The left-hand side is $2x + 5 = 2(5) + 5 = 10 + 5 = 15$.
- The right-hand side is $6x - 15 = 6(5) - 15 = 30 - 15 = 15$.

Since both sides equal 15, this shows that the answer $x = 5$ correctly solves the given equation.

Chapter 16 Problems

Directions: Isolate the unknown to solve each equation.

Problem 1. Solve $5x - 12 = 18$.

Problem 2. Solve $x + 9 = 17$.

Problem 3. Solve $8 + 3x = 7x$.

Problem 4. Solve $6x = 54$.

Problem 5. Solve $9 - x = 2x$.

Problem 6. Solve $4x = 28 - 3x$.

Problem 7. Solve $8x + 6 = 14$.

Problem 8. Solve $21 + 2x = 5x$.

Problem 9. Solve $7x - 28 = 3x$.

Problem 10. Solve $4 - x = 12$.

Problem 11. Solve $48 = 6x + 18$.

Problem 12. Solve $12x - 72 = 4x$.

Problem 13. Solve $6 + 5x = 2x$.

Problem 14. Solve $4x = 8 - 4x$.

Problem 15. Solve $16x = 9x - 56$.

Problem 16. Solve $30 = 6 - 8x$.

Problem 17. Solve $15 - x = 4x$.

Problem 18. Solve $14x + 90 = 5x$.

Problem 19. Solve $5x + 5 = 3x + 19$.

Problem 20. Solve $9x - 8 = 4x + 12$.

Problem 21. Solve $6x - 32 = 32 - 2x$.

Problem 22. Solve $16 + 7x = 36 - 3x$.

Problem 23. Solve $3 - 7x = 7 - 9x$.

Problem 24. Solve $3 - x = 5 + x$.

Problem 25. Solve $14 + 4x = 8x + 50$.

Problem 26. Solve $2x + 28 = 8x - 20$.

Problem 27. Solve $-2x + 18 = 60 - 8x$.

Problem 28. Solve $13x - 31 = 6x + 32$.

Problem 29. Solve $2x - 21 = 5x - 21$.

Problem 30. Solve $6 + 4x = 31 + 9x$.

Problem 31. Solve $44 - 2x = 12 - 10x$.

Problem 32. Solve $x + 12 = 9x - 12$.

Problem 33. Solve $-3 + 9x = 3x + 9$.

Problem 34. Solve $7x - 11 = 1 - 5x$.

Problem 35. Solve $4 + 20x = 100 + 12x$. **Problem 36**. Solve $32 - 8x = 41 - 7x$.

Problem 37. Solve $9x - 63 = 16x + 28$. **Problem 38**. Solve $22x + 150 - 16x = 30$.

Problem 39. Solve $-1 - x = 10x + 120$. **Problem 40**. Solve $-11x + 80 = 8 - 23x$.

Problem 41. Solve $4x - 27 = -2x + 9x$. **Problem 42**. Solve $100 - 7x = 8x - 50$.

Problem 43. Solve $6x - 4 + 2x - 7 + x = 8 + 9x - 9 - 3x - 1$.

Problem 44. Solve $14 - 12x - 35 + 8x - 28 = 5x - 63 + 9x + 70 - 11x$.

Problem 45. Solve $-24x - 96 + 36x + 144 - 48x = 60x - 108 - 72x - 60 + 84x$.

Combine Like Terms with Fractions

If one or more like terms involve fractions, such as $\frac{3x}{4} + \frac{2x}{3}$, the terms can be combined by adding (or subtracting, when there is a relative minus sign) fractions. Recall that the way to add or subtract fractions is to make a common denominator. For example:

$$\frac{3x}{4} + \frac{2x}{3} = \frac{9x}{12} + \frac{8x}{12} = \frac{17x}{12}$$

It may be helpful to review Chapter 1.

Chapter 17 Examples

Example 1. Simplify $\frac{x}{2} + \frac{2}{3} - \frac{x}{6} + \frac{3}{5}$.

Multiply $\frac{x}{2}$ by $\frac{3}{3}$ to make a common denominator of 6. Multiply $\frac{2}{3}$ by $\frac{5}{5}$ and multiply $\frac{3}{5}$ by $\frac{3}{3}$ to make a common denominator of 15. Divide $2x$ and 6 each by 2 to reduce $\frac{2x}{6}$ to $\frac{x}{3}$.

$$\frac{x}{2} + \frac{2}{3} - \frac{x}{6} + \frac{3}{5} = \left(\frac{3x}{6} - \frac{x}{6}\right) + \left(\frac{10}{15} + \frac{9}{15}\right) = \frac{2x}{6} + \frac{19}{15} = \frac{x}{3} + \frac{19}{15}$$

Example 2. Simplify $\frac{x^2}{3} - \frac{5x}{2} + \frac{x^2}{4} + \frac{7x}{4}$.

Multiply $\frac{x^2}{3}$ by $\frac{4}{4}$ and multiply $\frac{x^2}{4}$ by $\frac{3}{3}$ to make a common denominator of 12. Multiply $\frac{5x}{2}$ by $\frac{2}{2}$ to make a common denominator of 4.

$$\frac{x^2}{3} - \frac{5x}{2} + \frac{x^2}{4} + \frac{7x}{4} = \left(\frac{4x^2}{12} + \frac{3x^2}{12}\right) + \left(-\frac{10x}{4} + \frac{7x}{4}\right) = \frac{7x^2}{12} - \frac{3x}{4}$$

Chapter 17 Problems

Directions: Combine like terms to simplify each expression.

Problem 1. Simplify $\frac{x}{4} + \frac{2}{3} - \frac{x}{6} + \frac{1}{2}$.

Problem 2. Simplify $\frac{2x^2}{3} + \frac{5}{6} + \frac{4x^2}{9} - \frac{1}{3}$.

Problem 3. Simplify $x^2 + 2x - \frac{x^2}{5} - \frac{x}{4}$.

Problem 4. Simplify $\frac{7x}{4} - \frac{3}{2} + \frac{9x}{8} - \frac{2}{3}$.

Problem 5. Simplify $\frac{5x^2}{6} - \frac{1}{4} + \frac{2x^2}{9} + \frac{5}{2}$.

Problem 6. Simplify $\frac{x^3}{2} - \frac{x^3}{4} + \frac{x^2}{2} - \frac{x^2}{6}$.

Problem 7. Simplify $-\frac{2x}{7} - \frac{7}{6} + \frac{5x}{2} + \frac{1}{6}$.

Problem 8. Simplify $\frac{11x^2}{12} + 3x - \frac{x^2}{4} - \frac{4x}{3}$.

Problem 9. Simplify $\frac{x^2}{2} + \frac{x^3}{6} - \frac{x^2}{4} + \frac{x^3}{3}$.

Problem 10. Simplify $\frac{x^2}{8} + 2 + \frac{x^2}{6} - 9$.

Problem 11. Simplify $\frac{3x}{4} + \frac{x}{6} + \frac{x}{12}$.

Problem 12. Simplify $7x^2 - \frac{3}{8} - 2x^2 + \frac{5}{8}$.

Problem 13. Simplify $\frac{x^3}{12} - \frac{x}{18} - \frac{x^3}{6} - \frac{x}{6}$.

Problem 14. Simplify $x^2 + \frac{3x}{8} - \frac{4x^2}{7} - 2x$.

Problem 15. Simplify $\frac{8x^5}{9} - \frac{7x^2}{8} - \frac{4x^5}{3} + \frac{3x^2}{8}$.

Problem 16. Simplify $-\frac{8x}{15} + 6 - \frac{7x}{10} - \frac{9}{2}$.

Problem 17. Simplify $\frac{5}{4} + \frac{2x^2}{5} - \frac{9}{10} - \frac{x^2}{10}$.

Problem 18. Simplify $\frac{6x^4}{11} - \frac{9x}{4} + \frac{11x^4}{6} - \frac{2x}{3}$.

Problem 19. Simplify $\frac{7x}{12} - \frac{5x}{24} + \frac{11x}{18}$.

Problem 20. Simplify $\frac{5x^2}{36} - \frac{5}{18} - \frac{7x^2}{24} - \frac{7}{6}$.

Isolate the Unknown with Fractions

The problems in this chapter can be solved by isolating the unknown (it may help to review Chapter 16). What is different about this chapter is that the coefficients or the answers (or both) are fractions. Recall the following points regarding fractions:

- To add or subtract fractions, first make a common denominator (Chapter 17).
- To divide by a fraction, multiply by its reciprocal. (The **reciprocal** of a fraction has the numerator and denominator swapped. For example, $\frac{3}{2}$ is the reciprocal of $\frac{2}{3}$.) For example, $\frac{8}{9} \div \frac{2}{3} = \frac{8}{9}\left(\frac{3}{2}\right) = \frac{24}{18} = \frac{24 \div 6}{18 \div 6} = \frac{4}{3}$.

Chapter 18 Examples

Example 1. Solve the equation $\frac{2}{3}x = \frac{5}{8} + \frac{x}{4}$.

Method 1: Bring $\frac{x}{4}$ to the left side. When $\frac{x}{4}$ moves to the left-hand side, it changes sign. (An equivalent way to describe this is: "Subtract $\frac{x}{4}$ from both sides of the equation.")

$$\frac{2}{3}x - \frac{x}{4} = \frac{5}{8}$$

On the left-hand side, combine like terms (Chapter 17): $\frac{8x}{12} - \frac{3x}{12} = \frac{5x}{12}$.

$$\frac{5x}{12} = \frac{5}{8}$$

Multiply both sides of the equation by $\frac{12}{5}$. Why? Dividing by $\frac{5}{12}$ equates to multiplying by $\frac{12}{5}$. The main idea behind this is that $\frac{5x}{12}\left(\frac{12}{5}\right) = \frac{60x}{60} = x$.

$$\frac{5x}{12}\left(\frac{12}{5}\right) = \frac{5}{8}\left(\frac{12}{5}\right)$$

The left-hand side simplifies to $\frac{60x}{60} = x$. The right-hand side is $\frac{60}{40} = \frac{3}{2}$.

$$x = \frac{3}{2}$$

Method 2: Multiply every term of $\frac{2}{3}x = \frac{5}{8} + \frac{x}{4}$ by 24. Why? The least common multiple (LCM) of the three denominators (3, 8, and 4) equals 24. Multiplying both sides of the equation by 24 will effectively remove all of the fractions from the equation. Note that $\frac{2}{3}(24) = \frac{48}{3} = 16, \frac{5}{8}(24) = \frac{120}{8} = 15,$ and $\frac{1}{4}(24) = 6$.

$$16x = 15 + 6x$$
$$10x = 15$$
$$x = \frac{15}{10} = \frac{3}{2}$$

Check the answer: Plug $x = \frac{3}{2}$ into each side of the original equation. Verify that both sides of the equation equal the same value.

- The left-hand side is $\frac{2x}{3} = \frac{2}{3}\left(\frac{3}{2}\right) = \frac{6}{6} = 1$.
- The right-hand side is $\frac{5}{8} + \frac{x}{4} = \frac{5}{8} + \frac{1}{4}\left(\frac{3}{2}\right) = \frac{5}{8} + \frac{3}{8} = \frac{8}{8} = 1$.

Since both sides equal 1, this shows that the answer $x = \frac{3}{2}$ correctly solves the given equation.

Example 2. Solve the equation $\frac{2}{3} + \frac{x}{2} = 2$.

Method 1: Bring $\frac{2}{3}$ to the left side. When $\frac{2}{3}$ moves to the left-hand side, it changes sign. (An equivalent way to describe this is: "Subtract $\frac{2}{3}$ from both sides of the equation.")

$$\frac{x}{2} = 2 - \frac{2}{3}$$

On the right-hand side, combine like terms (Chapter 17): $\frac{6}{3} - \frac{2}{3} = \frac{4}{3}$.

$$\frac{x}{2} = \frac{4}{3}$$

Multiply both sides of the equation by 2. The main idea is that $\frac{x}{2}(2) = x$.

$$\frac{x}{2}(2) = \frac{4}{3}(2)$$

The left-hand side simplifies to $\frac{2x}{2} = x$. The right-hand side equals $\frac{8}{3}$.

$$x = \frac{8}{3}$$

Method 2: Multiply every term of $\frac{2}{3} + \frac{x}{2} = 2$ by 6. Why? The least common multiple (LCM) of the denominators (3 and 2) equals 6. Multiplying both sides of the equation by 6 will effectively remove all of the fractions from the equation. Note that $\frac{2}{3}(6) = \frac{12}{3} = 4$ and $\frac{x}{2}(6) = \frac{6x}{2} = 3x$.

$$4 + 3x = 12$$
$$3x = 8$$
$$x = \frac{8}{3}$$

Check the answer: Plug $x = \frac{8}{3}$ into each side of the original equation. Verify that both sides of the equation equal the same value.

- The left-hand side is $\frac{2}{3} + \frac{x}{2} = \frac{2}{3} + \frac{1}{2}\left(\frac{8}{3}\right) = \frac{2}{3} + \frac{8}{6} = \frac{2}{3} + \frac{4}{3} = \frac{6}{3} = 2$.
- The right-hand side is 2.

Since both sides equal 2, this shows that the answer $x = \frac{8}{3}$ correctly solves the given equation.

Chapter 18 Problems

Directions: Isolate the unknown to solve each equation.

Problem 1. Solve $x + \frac{2}{3} = \frac{3}{4}$.

Problem 2. Solve $\frac{5}{3} = \frac{x}{9}$.

Problem 3. Solve $\frac{5x}{6} = \frac{x}{3} + \frac{1}{18}$.

Problem 4. Solve $\frac{3x}{2} - \frac{1}{4} = \frac{17}{16}$.

Problem 5. Solve $6x = \frac{9}{8}$.

Problem 6. Solve $\frac{x}{4} = \frac{x}{6} + \frac{8}{3}$.

Problem 7. Solve $\frac{x}{2} - \frac{1}{4} = \frac{x}{3} + \frac{3}{8}$.

Problem 8. Solve $\frac{3x}{5} + \frac{8}{15} = \frac{5x}{6} + \frac{1}{2}$.

Problem 9. Solve $2 - \frac{x}{6} = \frac{2x}{9} + \frac{5}{18}$.

Problem 10. Solve $4 - \frac{7x}{5} = 1 + \frac{3x}{10}$.

Problem 11. Solve $\frac{2x}{3} - \frac{1}{12} = x - \frac{1}{3}$.

Problem 12. Solve $-\frac{x}{6} + \frac{5}{2} = \frac{x}{8} - \frac{7}{4}$.

Problem 13. Solve $\frac{3}{4} - x = \frac{1}{6} + \frac{x}{3}$.

Problem 14. Solve $\frac{7x}{9} - \frac{3}{8} = \frac{x}{12} + \frac{9}{10}$.

Problem 15. Solve $\dfrac{7}{12} + \dfrac{7x}{2} = \dfrac{5}{9} + \dfrac{9x}{4}$.

Problem 16. Solve $\dfrac{x}{7} - \dfrac{7}{3} = \dfrac{7}{6} - \dfrac{3x}{7}$.

Problem 17. Solve $\dfrac{8x}{15} + 3 = -\dfrac{7x}{6} + \dfrac{9}{4}$.

Problem 18. Solve $\dfrac{4}{9} - \dfrac{5x}{3} = \dfrac{7}{3} - x$.

Problem 19. Solve $\dfrac{1}{6} - \dfrac{x}{8} = \dfrac{1}{18} - \dfrac{x}{12}$.

Problem 20. Solve $\dfrac{8x}{5} + \dfrac{9}{4} = \dfrac{2x}{3} - \dfrac{11}{6}$.

Problem 21. Solve $\dfrac{8x}{9} - \dfrac{23}{18} = \dfrac{5x}{6} - \dfrac{7x}{12}$.

Problem 22. Solve $-\dfrac{7}{8} + \dfrac{3x}{8} = \dfrac{5}{6} - \dfrac{11x}{12}$.

Sum and Product Exercises

The problems in this chapter provide a product and a sum of two integers. The goal is to determine the two integers. This technique is useful for factoring simple quadratic expressions, which is the topic of Chapter 20.

Given the product and sum of two integers, one way to find the two integers is to make a list of the factors of the product. For example, suppose that the product is 30 and the sum is 17. The product, which is 30, can be factored as 1×30, 2×15, 3×10, or as 5×6. It can also be factored with two negative numbers: $(-1) \times (-30)$, $(-2) \times (-15)$, $(-3) \times (-10)$, or as $(-5) \times (-6)$. For which pair of factors does the sum of the factors equal the desired sum of 17? The two integers are 2 and 15 because $2 \times 15 = 30$ and because $2 + 15 = 17$.

If the product is positive, either both factors are positive or both factors are negative. If the product is negative, one factor is positive and the other factor is negative. For example, a product of 7 could be formed by 7×1 or by $(-7) \times (-1)$, which are either both positive or both negative. In contrast, a product of -7 could be formed by $(-7) \times 1$ or by $7 \times (-1)$, where one number is positive and the other is negative.

Note that $(-7) \times 1$ is different from $7 \times (-1)$ in an important regard: For $(-7) \times 1$ the sum of the two numbers is $-7 + 1 = -6$, whereas for $7 \times (-1)$ the sum of the two numbers is $7 + (-1) = 6$.

Chapter 19 Examples

Example 1. Which two numbers have a product of 20 and a sum of 9?

List the different ways that two integers can have a product of 20:

$$1 \times 20 \quad , \quad 2 \times 10 \quad , \quad 4 \times 5$$
$$(-1) \times (-20) \quad , \quad (-2) \times (-10) \quad , \quad (-4) \times (-5)$$

(It is not necessary to count 20×1 separately from 1×20, for example.) For which of these multiplications is the sum equal to 9? The answer is 4 and 5 because $4 \times 5 = 20$ and because $4 + 5 = 9$. **Tip**: <u>Both</u> numbers can't be negative if the sum is positive.

Example 2. Which two numbers have a product of -50 and a sum of 5?

List the different ways that two integers can have a product of -50:

$$1 \times (-50) \quad , \quad 2 \times (-25) \quad , \quad 5 \times (-10)$$
$$(-1) \times 50 \quad , \quad (-2) \times 25 \quad , \quad (-5) \times 10$$

For which of these multiplications is the sum equal to 5? The answer is -5 and 10 because $(-5) \times 10 = -50$ and because $-5 + 10 = 5$.

Example 3. Which two numbers have a product of 54 and a sum of -21?

List the different ways that two integers can have a product of 54:

$$1 \times 54 \quad , \quad 2 \times 27 \quad , \quad 3 \times 18 \quad , \quad 6 \times 9$$
$$(-1) \times (-54) \quad , \quad (-2) \times (-27) \quad , \quad (-3) \times (-18) \quad , \quad (-6) \times (-9)$$

For which of these multiplications is the sum equal to -21? The answer is -3 and -18 because $(-3) \times (-18) = 54$ and because $-3 + (-18) = -21$. **Tip**: <u>Both</u> numbers can't be positive if the sum is negative.

Example 4. Which two numbers have a product of -77 and a sum of -4?

List the different ways that two integers can have a product of -77:

$$1 \times (-77) \quad , \quad 7 \times (-11)$$
$$(-1) \times 77 \quad , \quad (-7) \times 11$$

For which of these multiplications is the sum equal to -4? The answer is 7 and -11 because $7 \times (-11) = -77$ and because $7 + (-11) = -4$.

Chapter 19 Problems

Directions: Find the numbers that make the indicated sum and product.

Problem 1. Which two numbers have a product of 6 and a sum of 5?

Problem 2. Which two numbers have a product of 8 and a sum of 6?

Problem 3. Which two numbers have a product of 24 and a sum of 10?

Problem 4. Which two numbers have a product of -15 and a sum of 2?

Problem 5. Which two numbers have a product of -20 and a sum of -1?

Problem 6. Which two numbers have a product of -25 and a sum of 0?

Problem 7. Which two numbers have a product of 18 and a sum of -9?

Problem 8. Which two numbers have a product of 100 and a sum of 25?

Problem 9. Which two numbers have a product of 48 and a sum of 16?

Problem 10. Which two numbers have a product of -72 and a sum of 21?

Problem 11. Which two numbers have a product of 60 and a sum of 17?

Problem 12. Which two numbers have a product of -60 and a sum of -17?

Problem 13. Which two numbers have a product of 36 and a sum of 13?

Problem 14. Which two numbers have a product of 24 and a sum of -11?

Problem 15. Which two numbers have a product of 144 and a sum of 30?

Factor Simple Quadratic Expressions

A **quadratic** expression has a term where the variable is squared (like $3x^2$) and may also have a linear term (like $4x$) or a constant term (like 15). An example of a quadratic expression is $3x^2 + 4x + 15$. This is a quadratic <u>expression</u>, NOT a quadratic equation, because it does not include an equal ($=$) sign.

The expression $x^2 - 7x + 10$ can be factored by finding two numbers that have a sum of -7 and a product of 10. The two numbers are -2 and -5 because $-2 + (-5) = -7$ and because $(-2)(-5) = 10$. These numbers can be used to factor $x^2 - 7x + 10$ as:
$$x^2 - 7x + 10 = (x - 2)(x - 5)$$
Check the answer using the FOIL method (Chapter 13):
$$(x - 2)(x - 5) = x(x) + x(-5) - 2(x) - 2(-5)$$
$$= x^2 - 5x - 2x + 10 = x^2 - 7x + 10$$
The problems in this chapter have the form $x^2 + bx + c$ and can be factored by finding two numbers with a sum equal to b and a product equal to c. (Not all quadratics can be factored so easily.) The strategy from Chapter 19 can be used to find the two numbers.

Chapter 20 Examples

Example 1. Factor $x^2 + 9x + 14$.
List the different ways that two integers can have a product of 14:
$$1 \times 14 \quad , \quad 2 \times 7$$
$$(-1) \times (-14) \quad , \quad (-2) \times (-7)$$
Since $2 + 7 = 9$, the expression $x^2 + 9x + 14$ can be factored as $(x + 2)(x + 7)$.

Check the answer: Use the FOIL method.

$$(x + 2)(x + 7) = x(x) + x(7) + 2(x) + 2(7)$$
$$= x^2 + 7x + 2x + 14 = x^2 + 9x + 14$$

Example 2. Factor $x^2 + 38x - 39$.

List the different ways that two integers can have a product of -39:

$$1 \times (-39) \quad , \quad 3 \times (-13)$$
$$(-1) \times 39 \quad , \quad (-3) \times 13$$

Since $-1 + 39 = 38$, the expression $x^2 + 38x - 39$ can be factored as $(x - 1)(x + 39)$.

Check the answer: Use the FOIL method.

$$(x - 1)(x + 39) = x(x) + x(39) - 1(x) - 1(39)$$
$$= x^2 + 39x - x - 39 = x^2 + 38x - 39$$

Example 3. Factor $x^2 - 6x + 8$.

List the different ways that two integers can have a product of 8:

$$1 \times 8 \quad , \quad 2 \times 4$$
$$(-1) \times (-8) \quad , \quad (-2) \times (-4)$$

Since $-2 + (-4) = -6$, the expression $x^2 - 6x + 8$ can be factored as $(x - 2)(x - 4)$.

Check the answer: Use the FOIL method.

$$(x - 2)(x - 4) = x(x) + x(-4) - 2(x) - 2(-4)$$
$$= x^2 - 4x - 2x + 8 = x^2 - 6x + 8$$

Example 4. Factor $x^2 - 7x - 30$.

List the different ways that two integers can have a product of -30:

$$1 \times (-30) \quad , \quad 2 \times (-15) \quad , \quad 3 \times (-10) \quad , \quad 6 \times (-5)$$
$$(-1) \times 30 \quad , \quad (-2) \times 15 \quad , \quad (-3) \times 10 \quad , \quad (-6) \times 5$$

Since $3 + (-10) = -7$, the expression $x^2 - 7x - 30$ can be factored as $(x + 3)(x - 10)$.

Check the answer: Use the FOIL method.

$$(x + 3)(x - 10) = x(x) + x(-10) + 3(x) + 3(-10)$$
$$= x^2 - 10x + 3x - 30 = x^2 - 7x - 30$$

Chapter 20 Problems

Directions: Factor each expression.

Problem 1. Factor $x^2 + 7x + 10$.

Problem 2. Factor $x^2 - 9x + 8$.

Problem 3. Factor $x^2 - 4x - 12$.

Problem 4. Factor $x^2 + 3x - 18$.

Problem 5. Factor $x^2 + 11x + 24$.

Problem 6. Factor $x^2 - 13x - 30$.

Problem 7. Factor $x^2 - 17x + 72$.

Problem 8. Factor $x^2 - 20x + 64$.

Problem 9. Factor $x^2 + 3x - 4$.

Problem 10. Factor $x^2 - 16x + 28$.

Problem 11. Factor $x^2 + 18x + 56$.

Problem 12. Factor $x^2 - x - 42$.

Problem 13. Factor $x^2 - 30x + 81$.

Problem 14. Factor $x^2 + 3x - 40$.

Problem 15. Factor $x^2 - 4x - 45$.

Problem 16. Factor $x^2 + 12x + 20$.

Problem 17. Factor $x^2 - 21x + 80$.

Problem 18. Factor $x^2 + 20x - 96$.

Answer Key

Chapter 1 Answers

Problem 1. $4x + 5 + 11x + 3 = (4x + 11x) + (5 + 3) = 15x + 8$

Problem 2. $2x^2 + 16x + x^2 - 6x = (2x^2 + 1x^2) + (16x - 6x) = 3x^2 + 10x$

Problem 3. $17x^2 - 10x^2 - 7 + 6 = (17x^2 - 10x^2) + (-7 + 6) = 7x^2 - 1$

Problem 4. $3x + 9 - 4x = (3x - 4x) + 9 = -1x + 9 = -x + 9$

Problem 5. $x^2 + 7 + x^2 + 8 = (1x^2 + 1x^2) + (7 + 8) = 2x^2 + 15$

Problem 6. $2x + 36x - 26x = (2x + 36x - 26x) = 12x$

Problem 7. $-12x - 3 + 6x - 14 = (-12x + 6x) + (-3 - 14) = -6x - 17$

Problem 8. $-7x^2 + 4x^2 - 6x^2 = (-7x^2 + 4x^2 - 6x^2) = -9x^2$

Problem 9. $27x^2 + 5x^3 + 11x^2 + 8x^3 = (5x^3 + 8x^3) + (27x^2 + 11x^2) = 13x^3 + 38x^2$

Problem 10. $6x + 3 + x - 7 + 2x = (6x + 1x + 2x) + (3 - 7) = 9x - 4$

Problem 11. $13 - 9x + 6 = -9x + (13 + 6) = -9x + 19$

Problem 12. $5x^2 - 8x^2 - 4x^2 = (5x^2 - 8x^2 - 4x^2) = -7x^2$

Problem 13. $17x^3 - 14x - 12x^3 - 14x = (17x^3 - 12x^3) + (-14x - 14x) = 5x^3 - 28x$

Problem 14. $16 + 13x^2 - 17 - 22x^2 - 6 = (13x^2 - 22x^2) + (16 - 17 - 6) = -9x^2 - 7$

Problem 15. $25x^3 + 28x^4 - 21x^4 - 15x^3 = (28x^4 - 21x^4) + (25x^3 - 15x^3) = 7x^4 + 10x^3$

Problem 16. $-8x - 17 + 4x - 10 = (-8x + 4x) + (-17 - 10) = -4x - 27$

Problem 17. $14 + 22x^2 - 19 + 21x^2 = (22x^2 + 21x^2) + (14 - 19) = 43x^2 - 5$

Problem 18. $6x - 4x + 9x - x = (6x - 4x + 9x - 1x) = 10x$

Problem 19. $25x^5 + 5x^3 + 50x^5 - 6x^3 = (25x^5 + 50x^5) + (5x^3 - 6x^3) = 75x^5 - x^3$

Problem 20. $3x^2 - 5 - 8x^2 + 4 - 7x^2 = (3x^2 - 8x^2 - 7x^2) + (-5 + 4) = -12x^2 - 1$

Problem 21. $-9x - 2x^3 + 4x + 10x^3 = (-2x^3 + 10x^3) + (-9x + 4x) = 8x^3 - 5x$

Problem 22. $6x + 12 - 4 - x - 9 = (6x - 1x) + (12 - 4 - 9) = 5x - 1$

Problem 23. $19x^2 + 13x^3 - 2x^2 - 3x^3 = (13x^3 - 3x^3) + (19x^2 - 2x^2) = 10x^3 + 17x^2$

Problem 24. $15x + 21 - 12x + 4 = (15x - 12x) + (21 + 4) = 3x + 25$

Problem 25. $4x^2 - 9x^2 + 13 + 7x^2 + 18 = (4x^2 - 9x^2 + 7x^2) + (13 + 18) = 2x^2 + 31$

Problem 26. $6x^3 - 5x^5 + 2x^3 - 2x^5 = (-5x^5 - 2x^5) + (6x^3 + 2x^3) = -7x^5 + 8x^3$

Problem 27. $41x + 10x^2 + 33x + 9x^2 + 8x = (10x^2 + 9x^2) + (41x + 33x + 8x) = 19x^2 + 82x$

Problem 28. $26 - 9x^3 - 26 - 4x^3 = (-9x^3 - 4x^3) + (26 - 26) = -13x^3 + 0 = -13x^3$

Problem 29. $13x^2 + 8 - 14 - 14x^2 + 2 = (13x^2 - 14x^2) + (8 - 14 + 2) = -x^2 - 4$

Problem 30. $6x + 12 + 6x - 1 + 3x - 4 = (6x + 6x + 3x) + (12 - 1 - 4) = 15x + 7$

Problem 31. $x^3 + 25x^2 - 6x^3 - 22x^2 + 9x^3 = (1x^3 - 6x^3 + 9x^3) + (25x^2 - 22x^2) =$
$4x^3 + 3x^2$

Problem 32. $10x - 1 - 4x + 5x - 2 - 3 = (10x - 4x + 5x) + (-1 - 2 - 3) = 11x - 6$

Problem 33. $-44x + 6x^2 - x + 50x - 7x^2 = (6x^2 - 7x^2) + (-44x - 1x + 50x) = -x^2 + 5x$

Problem 34. $15 + 3x^2 - 16 - 2x^2 + 1 - x^2 = (3x^2 - 2x^2 - 1x^2) + (15 - 16 + 1) = 0$

Problem 35. $6x^3 - 8x^3 - 11x^4 - 16x^4 - 11x^3 = (-11x^4 - 16x^4) + (6x^3 - 8x^3 - 11x^3) =$
$-27x^4 - 13x^3$

Problem 36. $-x - 1 - x - 1 + 3x - 1 = (-1x - 1x + 3x) + (-1 - 1 - 1) = 1x - 3 = x - 3$

Problem 37. $-3x^5 - 4x^5 + 4x^5 - 11 - 3 = (-3x^5 - 4x^5 + 4x^5) + (-11 - 3) = -3x^5 - 14$

Problem 38. $21x^2 + 7x + 9x^2 + 14x + 10x = (21x^2 + 9x^2) + (7x + 14x + 10x) =$
$30x^2 + 31x$

Problem 39. $6x - 12x^3 + 11x^3 - 3x - x^3 + 4x =$
$(-12x^3 + 11x^3 - 1x^3) + (6x - 3x + 4x) = -2x^3 + 7x$

Problem 40. $x^2 + 2x + 4 + 2x^2 + 2x + 2 = (1x^2 + 2x^2) + (2x + 2x) + (4 + 2) =$
$3x^2 + 4x + 6$

Problem 41. $90x + x^4 - 20x - 70x - 23x^4 = (x^4 - 23x^4) + (90x - 20x - 70x) =$
$-22x^4 + 0 = -22x^4$

Problem 42. $x^5 + 8x^2 - x^5 + 17x^2 - 2x^5 - 10x^2 =$
$(x^5 - x^5 - 2x^5) + (8x^2 + 17x^2 - 10x^2) = -2x^5 + 15x^2$

Problem 43. $9 - 4x - 21 + 6x - 11 + 8x = (-4x + 6x + 8x) + (9 - 21 - 11) = 10x - 23$

Problem 44. $8 - 7x^2 - 3 + 6x^2 - 1 = (-7x^2 + 6x^2) + (8 - 3 - 1) = -1x^2 + 4 = -x^2 + 4$

Problem 45. $6x^3 + 22x - 8x^3 + 11x - 6x^3 + 4x =$
$(6x^3 - 8x^3 - 6x^3) + (22x + 11x + 4x) = -8x^3 + 37x$

Problem 46. $-9x^4 - 2x^2 + 4x^4 + 11x^2 - 3x^2 = (-9x^4 + 4x^4) + (-2x^2 + 11x^2 - 3x^2) =$
$-5x^4 + 6x^2$

Problem 47. $x - 12 - x + 3x - 4 - 1 = (1x - 1x + 3x) + (-12 - 4 - 1) = 3x - 17$

Problem 48. $6x^3 - x^2 + 5x^3 + 6x^3 - 3x^2 = (6x^3 + 5x^3 + 6x^3) + (-1x^2 - 3x^2) = 17x^3 - 4x^2$

Problem 49. $-5x^2 - 4x^2 + 8 - 7x^2 + 9 + 10 = (-5x^2 - 4x^2 - 7x^2) + (8 + 9 + 10) =$
$-16x^2 + 27$

Problem 50. $16x + 15 + 4x + 8x + 22 + 4 = (16x + 4x + 8x) + (15 + 22 + 4) = 28x + 41$

Problem 51. $5 - 11x^3 + x + 6 + 6x^3 + 10 = (-11x^3 + 6x^3) + (1x) + (5 + 6 + 10) =$
$-5x^3 + x + 21$

Problem 52. $5x^2 - 6x + 3x - 2x^2 - 7x = (5x^2 - 2x^2) + (-6x + 3x - 7x) = 3x^2 - 10x$

Problem 53. $2x^3 - 2x - 14x - 3x^3 - 20x^3 + 14x =$
$(2x^3 - 3x^3 - 20x^3) + (-2x - 14x + 14x) = -21x^3 - 2x$

Problem 54. $18x + 27x + 5x^2 + 10x^2 + 31x + 22x =$
$(5x^2 + 10x^2) + (18x + 27x + 31x + 22x) = 15x^2 + 98x$

Problem 55. $-4x^3 + 4x^2 + 7x + 5x^3 - 5x^2 - 8x =$
$(-4x^3 + 5x^3) + (4x^2 - 5x^2) + (7x - 8x) = 1x^3 - 1x^2 - 1x = x^3 - x^2 - x$

Problem 56. $11x - 4 - 2x + 3x - 2 - 5 = (11x - 2x + 3x) + (-4 - 2 - 5) = 12x - 11$

Problem 57. $-11x^7 - 11x^4 - 11x^7 - 11x^7 - x^4 - 11x^4 =$
$(-11x^7 - 11x^7 - 11x^7) + (-11x^4 - 1x^4 - 11x^4) = -33x^7 - 23x^4$

Problem 58. $65x + 46 - x - 12x - 13 + 20 = (65x - 1x - 12x) + (46 - 13 + 20) = 52x + 53$

Problem 59. $43x^2 + 4 - 13x^2 + 112 - 12x^2 - 43 =$
$(43x^2 - 13x^2 - 12x^2) + (4 + 112 - 43) = 18x^2 + 73$

Problem 60. $5x^2 - 7x + x^3 - 7x + 2x^3 + 5x^2 = (x^3 + 2x^3) + (5x^2 + 5x^2) + (-7x - 7x) =$
$3x^3 + 10x^2 - 14x$

Problem 61. $47x^2 - 4 - 6x - x + 11 - 22x^2 =$
$(47x^2 - 22x^2) + (-6x - 1x) + (-4 + 11) = 25x^2 - 7x + 7$

Problem 62. $16 - 9x^3 + 4x - 11 - 11x^3 + 2x = (-9x^3 - 11x^3) + (4x + 2x) +$
$(16 - 11) = -20x^3 + 6x + 5$

Problem 63. $75 - 5x^3 - 45 - x^3 + 3 - 14 = (-5x^3 - 1x^3) + (75 - 45 + 3 - 14) = -6x^3 + 19$

Problem 64. $2x^2 + x + 6 + x^2 + 6x + 16 = (2x^2 + 1x^2) + (1x + 6x) + (6 + 16) =$
$3x^2 + 7x + 22$

Problem 65. $45x^3 + x^2 - 11x - 25x^3 - 55x^2 - 20x =$
$(45x^3 - 25x^3) + (1x^2 - 55x^2) + (-11x - 20x) = 20x^3 - 54x^2 - 31x$

Problem 66. $28 - 4x + 12 + 6x^2 - 1 + 8x - 4 =$
$(6x^2) + (-4x + 8x) + (28 + 12 - 1 - 4) = 6x^2 + 4x + 35$

Problem 67. $-25x^3 - x - 110 - 15x^3 + 15x + 25 =$
$(-25x^3 - 15x^3) + (-1x + 15x) + (-110 + 25) = -40x^3 + 14x - 85$

Problem 68. $8x^2 - 22 - 31 - 8x^2 - 8x^2 + 4 = (8x^2 - 8x^2 - 8x^2) + (-22 - 31 + 4) =$
$-8x^2 - 49$

Problem 69. $20 + 14x^2 - 22 - 3x^3 + 3x^2 - 2x^3 =$
$(-3x^3 - 2x^3) + (14x^2 + 3x^2) + (20 - 22) = -5x^3 + 17x^2 - 2$

Problem 70. $14 + x - 13x^2 - 5x + 7 - 4x^2 = (-13x^2 - 4x^2) + (1x - 5x) + (14 + 7) =$
$-17x^2 - 4x + 21$

Chapter 2 Answers

Problem 1. $x^2 x^4 = x^{2+4} = x^6$

Problem 2. $x^4 x = x^4 x^1 = x^{4+1} = x^5$

Problem 3. $2x^5 x^3 = 2x^{5+3} = 2x^8$

Problem 4. $x^2 x = x^2 x^1 = x^{2+1} = x^3$

Problem 5. $3x^4 x^5 = 3x^{4+5} = 3x^9$

Problem 6. $x^3 x^3 = x^{3+3} = x^6$

Problem 7. $xx^7 = x^1 x^7 = x^{1+7} = x^8$

Problem 8. $8x^7 x^6 = 8x^{7+6} = 8x^{13}$

Problem 9. $x^6 x^8 = x^{6+8} = x^{14}$

Problem 10. $6x^4 x^4 = 6x^{4+4} = 6x^8$

Problem 11. $x^8 x^9 = x^{8+9} = x^{17}$

Problem 12. $10x^5 x^5 = 10x^{5+5} = 10x^{10}$

Problem 13. $x^3 x^4 x^5 = x^{3+4+5} = x^{12}$

Problem 14. $x^7 x^8 x = x^7 x^8 x^1 = x^{7+8+1} = x^{16}$

Problem 15. $6x^2 x^2 x^2 = 6x^{2+2+2} = 6x^6$

Problem 16. $4x^5 xx^3 = 4x^5 x^1 x^3 = 4x^{5+1+3} = 4x^9$

Problem 17. $x^9 x^8 x^6 = x^{9+8+6} = x^{23}$

Problem 18. $9xx^8 x = 9x^1 x^8 x^1 = 9x^{1+8+1} = 9x^{10}$

Problem 19. $5x^4 x^3 x^2 x^2 = 5x^{4+3+2+2} = 5x^{11}$

Problem 20. $x^5 x^7 x^4 x = x^5 x^7 x^4 x^1 = x^{5+7+4+1} = x^{17}$

Problem 21. $(x^5)^2 = x^{(5)(2)} = x^{10}$

Problem 22. $(x^2)^4 = x^{(2)(4)} = x^8$

Problem 23. $(x^3)^3 = x^{(3)(3)} = x^9$

Problem 24. $(x^7)^5 = x^{(7)(5)} = x^{35}$

Problem 25. $(x^4)^8 = x^{(4)(8)} = x^{32}$

Problem 26. $(x^9)^6 = x^{(9)(6)} = x^{54}$

Problem 27. $(3x)^3 = 3^3 x^3 = 27x^3$

Problem 28. $(6x)^2 = 6^2 x^2 = 36x^2$

Problem 29. $(2x)^5 = 2^5 x^5 = 32x^5$

Problem 30. $(5x)^3 = 5^3 x^3 = 125x^3$

Problem 31. $(7x)^2 = 7^2 x^2 = 49x^2$

Problem 32. $(3x)^4 = 3^4 x^4 = 81x^4$

Problem 33. $(5x^3)^2 = 5^2 x^{(3)(2)} = 25x^6$

Problem 34. $(2x^4)^3 = 2^3 x^{(4)(3)} = 8x^{12}$

Problem 35. $(10x^5)^4 = 10^4 x^{(5)(4)} = 10{,}000x^{20}$

Problem 36. $(4x^3)^3 = 4^3 x^{(3)(3)} = 64x^9$

Problem 37. $(8x^4)^2 = 8^2 x^{(4)(2)} = 64x^8$

Problem 38. $(2x^9)^4 = 2^4 x^{(9)(4)} = 16x^{36}$

Problem 39. $(5x)^2(4x^2)^3 = 5^2 x^2 4^3 x^{(2)(3)} = 25x^2 64x^6 = 1600x^{2+6} = 1600x^8$

Problem 40. $(x^2)^4(6x^5)^2 = x^{(2)(4)} 6^2 x^{(5)(2)} = x^8 36x^{10} = 36x^{8+10} = 36x^{18}$

Chapter 3 Answers

Problem 1. $9(x + 6) = 9(x) + 9(6) = 9x + 54$

Problem 2. $x(x - 4) = x(x) + x(-4) = x^2 - 4x$

Problem 3. $3x(-x^2 + 7x) = 3x(-x^2) + 3x(7x) = -3x^3 + 21x^2$

Problem 4. $6x^3(2x + 5) = 6x^3(2x) + 6x^3(5) = 12x^4 + 30x^3$

Problem 5. $4x^2(-5x^2 - 3) = 4x^2(-5x^2) + 4x^2(-3) = -20x^4 - 12x^2$

Problem 6. $x^4(x^2 - 8x) = x^4(x^2) + x^4(-8x) = x^6 - 8x^5$

Problem 7. $2x(4x^3 + 6x^2) = 2x(4x^3) + 2x(6x^2) = 8x^4 + 12x^3$

Problem 8. $x^2(-x^2 + 5) = x^2(-x^2) + x^2(5) = -x^4 + 5x^2$

Problem 9. $4(7x^3 + x) = 4(7x^3) + 4(x) = 28x^3 + 4x$

Problem 10. $3x^2(-6x^4 - 5x^2) = 3x^2(-6x^4) + 3x^2(-5x^2) = -18x^6 - 15x^4$

Problem 11. $x(x + 9) = x(x) + x(9) = x^2 + 9x$

Problem 12. $8x^3(-6x^3 + 7x) = 8x^3(-6x^3) + 8x^3(7x) = -48x^6 + 56x^4$

Problem 13. $5x(x - 2) = 5x(x) + 5x(-2) = 5x^2 - 10x$

Problem 14. $2x(-3x^2 + 6x) = 2x(-3x^2) + 2x(6x) = -6x^3 + 12x^2$

Problem 15. $8(3x^3 + 7x^2) = 8(3x^3) + 8(7x^2) = 24x^3 + 56x^2$

Problem 16. $x(-x^4 - x) = x(-x^4) + x(-x) = -x^5 - x^2$

Problem 17. $5x^2(6x^3 + 5) = 5x^2(6x^3) + 5x^2(5) = 30x^5 + 25x^2$

Problem 18. $2x(-x^2 - 7x) = 2x(-x^2) + 2x(-7x) = -2x^3 - 14x^2$

Problem 19. $7(3x^4 - 6x^2) = 7(3x^4) + 7(-6x^2) = 21x^4 - 42x^2$

Problem 20. $6x^3(-5x^2 + 9x) = 6x^3(-5x^2) + 6x^3(9x) = -30x^5 + 54x^4$

Problem 21. $4x^2(8x^2 - 3) = 4x^2(8x^2) + 4x^2(-3) = 32x^4 - 12x^2$

Problem 22. $5x^3(-7x^6 - 4x^2) = 5x^3(-7x^6) + 5x^3(-4x^2) = -35x^9 - 20x^5$

Problem 23. $x^2(-x^3 + x) = x^2(-x^3) + x^2(x) = -x^5 + x^3$

Problem 24. $9x(5x^2 + 2x) = 9x(5x^2) + 9x(2x) = 45x^3 + 18x^2$

Problem 25. $7x^2(-3x^3 + 4x^2) = 7x^2(-3x^3) + 7x^2(4x^2) = -21x^5 + 28x^4$

Problem 26. $8x^4(7x^5 - 6x^3) = 8x^4(7x^5) + 8x^4(-6x^3) = 56x^9 - 48x^7$

Problem 27. $6x^7(6x^4 + 3x^2) = 6x^7(6x^4) + 6x^7(3x^2) = 36x^{11} + 18x^9$

Problem 28. $4x^5(-x^5 + x^3) = 4x^5(-x^5) + 4x^5(x^3) = -4x^{10} + 4x^8$

Problem 29. $9x^8(6x^6 - 9) = 9x^8(6x^6) + 9x^8(-9) = 54x^{14} - 81x^8$

Problem 30. $5x^6(-7x^9 - 8x^5) = 5x^6(-7x^9) + 5x^6(-8x^5) = -35x^{15} - 40x^{11}$

Chapter 4 Answers

Problem 1. $3(7x^2 + 2x + 4) = 3(7x^2) + 3(2x) + 3(4) = 21x^2 + 6x + 12$

Problem 2. $5x(4x^2 + 8x - 9) = 5x(4x^2) + 5x(8x) + 5x(-9) = 20x^3 + 40x^2 - 45x$

Problem 3. $2(5x^3 - 3x^2 + 6x) = 2(5x^3) + 2(-3x^2) + 2(6x) = 10x^3 - 6x^2 + 12x$

Problem 4. $x^2(-2x^2 - 4x + 6) = x^2(-2x^2) + x^2(-4x) + x^2(6) = -2x^4 - 4x^3 + 6x^2$

Problem 5. $4x(3x^4 - 6x^2 - 7) = 4x(3x^4) + 4x(-6x^2) + 4x(-7) = 12x^5 - 24x^3 - 28x$

Problem 6. $8(2x^2 + 3x + 9) = 8(2x^2) + 8(3x) + 8(9) = 16x^2 + 24x + 72$

Problem 7. $3x^2(7x^2 + 6x - 4) = 3x^2(7x^2) + 3x^2(6x) + 3x^2(-4) = 21x^4 + 18x^3 - 12x^2$

Problem 8. $7x(-4x^5 + 5x^3 - 8x) = 7x(-4x^5) + 7x(5x^3) + 7x(-8x) =$
$-28x^6 + 35x^4 - 56x^2$

Problem 9. $9(6x^3 + x^2 - 2x) = 9(6x^3) + 9(x^2) + 9(-2x) = 54x^3 + 9x^2 - 18x$

Problem 10. $2x^4(-3x^2 + 3x + 9) = 2x^4(-3x^2) + 2x^4(3x) + 2x^4(9) = -6x^6 + 6x^5 + 18x^4$

Problem 11. $x(x^3 - x^2 + 1) = x(x^3) + x(-x^2) + x(1) = x^4 - x^3 + x$

Problem 12. $5x^2(6x^5 - 4x^4 + 7) = 5x^2(6x^5) + 5x^2(-4x^4) + 5x^2(7) =$
$30x^7 - 20x^6 + 35x^2$

Problem 13. $6(x^4 + 9x^2 + 8) = 6(x^4) + 6(9x^2) + 6(8) = 6x^4 + 54x^2 + 48$

Problem 14. $4x^3(-6x^6 - 5x^3 - 7x^2) = 4x^3(-6x^6) + 4x^3(-5x^3) + 4x^3(-7x^2) =$
$-24x^9 - 20x^6 - 28x^5$

Problem 15. $8x(4x^5 + x^3 + 3) = 8x(4x^5) + 8x(x^3) + 8x(3) = 32x^6 + 8x^4 + 24x$

Problem 16. $x^5(9x^6 - 2x^2 + 3x) = x^5(9x^6) + x^5(-2x^2) + x^5(3x) = 9x^{11} - 2x^7 + 3x^6$

Problem 17. $6x^2(3x^2 + 4x + 5) = 6x^2(3x^2) + 6x^2(4x) + 6x^2(5) = 18x^4 + 24x^3 + 30x^2$

Problem 18. $2x(2x^3 + 8x - 9) = 2x(2x^3) + 2x(8x) + 2x(-9) = 4x^4 + 16x^2 - 18x$

Problem 19. $9x^4(-7x^5 - 6x^3 + 7x) = 9x^4(-7x^5) + 9x^4(-6x^3) + 9x^4(7x) =$
$-63x^9 - 54x^7 + 63x^5$

Problem 20. $7(4x^2 + 9x - 4) = 7(4x^2) + 7(9x) + 7(-4) = 28x^2 + 63x - 28$

Problem 21. $6x^3(-5x^7 - 3x^5 + 8x^4) = 6x^3(-5x^7) + 6x^3(-3x^5) + 6x^3(8x^4) =$
$-30x^{10} - 18x^8 + 48x^7$

Problem 22. $9x(x^3 + 6x^2 + 3) = 9x(x^3) + 9x(6x^2) + 9x(3) = 9x^4 + 54x^3 + 27x$

Problem 23. $x^3(-8x^5 - 2x^4 - x^2) = x^3(-8x^5) + x^3(-2x^4) + x^3(-x^2) = -8x^8 - 2x^7 - x^5$

Problem 24. $5(3x^6 - 9x^4 + 7x^2) = 5(3x^6) + 5(-9x^4) + 5(7x^2) = 15x^6 - 45x^4 + 35x^2$

Problem 25. $8x^2(-4x^4 - x^3 - 5) = 8x^2(-4x^4) + 8x^2(-x^3) + 8x^2(-5) =$
$-32x^6 - 8x^5 - 40x^2$

Problem 26. $5x^3(5x^2 + 7x - 9) = 5x^3(5x^2) + 5x^3(7x) + 5x^3(-9) = 25x^5 + 35x^4 - 45x^3$

Problem 27. $6x^5(3x^3 - 8x^2 + 7x) = 6x^5(3x^3) + 6x^5(-8x^2) + 6x^5(7x) =$
$18x^8 - 48x^7 + 42x^6$

Problem 28. $2x^4(-4x^8 + 5x^5 - x^2) = 2x^4(-4x^8) + 2x^4(5x^5) + 2x^4(-x^2) =$
$-8x^{12} + 10x^9 - 2x^6$

Problem 29. $6x(9x^5 - 7x^3 - 5x) = 6x(9x^5) + 6x(-7x^3) + 6x(-5x) =$
$54x^6 - 42x^4 - 30x^2$

Problem 30. $3x^7(2x^2 + 8x + 4) = 3x^7(2x^2) + 3x^7(8x) + 3x^7(4) = 6x^9 + 24x^8 + 12x^7$

Problem 31. $7x^2(-x^4 - 3x^2 + 7) = 7x^2(-x^4) + 7x^2(-3x^2) + 7x^2(7) =$
$-7x^6 - 21x^4 + 49x^2$

Problem 32. $5x^8(4x^6 + 9x^4 - 6) = 5x^8(4x^6) + 5x^8(9x^4) + 5x^8(-6) =$
$20x^{14} + 45x^{12} - 30x^8$

Problem 33. $8x^3(5x^9 - 8x^6 + 2x^3) = 8x^3(5x^9) + 8x^3(-8x^6) + 8x^3(2x^3) =$
$40x^{12} - 64x^9 + 16x^6$

Problem 34. $9x^6(3x^8 - 9x^5 - 4x^2) = 9x^6(3x^8) + 9x^6(-9x^5) + 9x^6(-4x^2) =$
$27x^{14} - 81x^{11} - 36x^8$

Chapter 5 Answers

Problem 1. $5(5x^3 + x^2 + 6x + 4) = 5(5x^3) + 5(x^2) + 5(6x) + 5(4) = 25x^3 + 5x^2 + 30x + 20$

Problem 2. $3x(6x^7 - 4x^5 + 3x^3 - 2x) = 3x(6x^7) + 3x(-4x^5) + 3x(3x^3) + 3x(-2x) =$
$18x^8 - 12x^6 + 9x^4 - 6x^2$

Problem 3. $x^3(4x^7 + 8x^5 + 6x^3 - 3x) = x^3(4x^7) + x^3(8x^5) + x^3(6x^3) + x^3(-3x) =$
$4x^{10} + 8x^8 + 6x^6 - 3x^4$

Problem 4. $2x^2(-9x^4 - 6x^3 + 3x^2 + x) =$
$2x^2(-9x^4) + 2x^2(-6x^3) + 2x^2(3x^2) + 2x^2(x) = -18x^6 - 12x^5 + 6x^4 + 2x^3$

Problem 5. $8(4x^6 - 6x^4 - 7x^2 + 9) = 8(4x^6) + 8(-6x^4) + 8(-7x^2) + 8(9) =$
$32x^6 - 48x^4 - 56x^2 + 72$

Problem 6. $9x^3(x^3 + 3x^2 - 6x - 9) = 9x^3(x^3) + 9x^3(3x^2) + 9x^3(-6x) + 9x^3(-9) =$
$9x^6 + 27x^5 - 54x^4 - 81x^3$

Problem 7. $6x^5(-5x^6 + 3x^3 + 4x^2 + 5x) =$
$6x^5(-5x^6) + 6x^5(3x^3) + 6x^5(4x^2) + 6x^5(5x) = -30x^{11} + 18x^8 + 24x^7 + 30x^6$

Problem 8. $4x(7x^3 + 6x^2 + 5x + 4) = 4x(7x^3) + 4x(6x^2) + 4x(5x) + 4x(4) =$
$28x^4 + 24x^3 + 20x^2 + 16x$

Problem 9. $5x^4(7x^9 - 8x^7 - 6x^5 - 3x^3) =$
$5x^4(7x^9) + 5x^4(-8x^7) + 5x^4(-6x^5) + 5x^4(-3x^3) = 35x^{13} - 40x^{11} - 30x^9 - 15x^7$

Problem 10. $7x^6(2x^7 - 4x^6 + 6x^4 - 10) =$
$7x^6(2x^7) + 7x^6(-4x^6) + 7x^6(6x^4) + 7x^6(-10) = 14x^{13} - 28x^{12} + 42x^{10} - 70x^6$

Problem 11. $4x^7(-9x^6 + 6x^5 - 9x^4 + x^3) =$
$4x^7(-9x^6) + 4x^7(6x^5) + 4x^7(-9x^4) + 4x^7(x^3) = -36x^{13} + 24x^{12} - 36x^{11} + 4x^{10}$

Problem 12. $9x^4(-7x^7 - 6x^5 - 5x^4 - 7x^2) =$
$9x^4(-7x^7) + 9x^4(-6x^5) + 9x^4(-5x^4) + 9x^4(-7x^2) = -63x^{11} - 54x^9 - 45x^8 - 63x^6$

Problem 13. $6x(-8x^9 - 2x^6 - x^5 + 100) =$
$6x(-8x^9) + 6x(-2x^6) + 6x(-x^5) + 6x(100) = -48x^{10} - 12x^7 - 6x^6 + 600x$

Problem 14. $4x^3(-6x^9 + 4x^8 - 9x^6 - 5x^3 - 7x) =$
$4x^3(-6x^9) + 4x^3(4x^8) + 4x^3(-9x^6) + 4x^3(-5x^3) + 4x^3(-7x) =$
$-24x^{12} + 16x^{11} - 36x^9 - 20x^6 - 28x^4$

Problem 15. $4x^4(7x^7 - 6x^6 - 5x^5 - 4x^4 + 8x^3) =$
$4x^4(7x^7) + 4x^4(-6x^6) + 4x^4(-5x^5) + 4x^4(-4x^4) + 4x^4(8x^3) =$
$28x^{11} - 24x^{10} - 20x^9 - 16x^8 + 32x^7$

Problem 16. $10x^7(-8x^7 + 4x^6 + 10x^5 - 9x^2 - 12x - 5) =$

$10x^7(-8x^7) + 10x^7(4x^6) + 10x^7(10x^5) + 10x^7(-9x^2) + 10x^7(-12x) + 10x^7(-5) =$

$-80x^{14} + 40x^{13} + 100x^{12} - 90x^9 - 120x^8 - 50x^7$

Problem 17. $8x^9(2x^{13} + 6x^{11} - 3x^9 - 7x^7 + 9x^5 - 4x^3 + 8x) =$

$8x^9(2x^{13}) + 8x^9(6x^{11}) + 8x^9(-3x^9) + 8x^9(-7x^7) + 8x^9(9x^5) + 8x^9(-4x^3) + 8x^9(8x) =$

$16x^{22} + 48x^{20} - 24x^{18} - 56x^{16} + 72x^{14} - 32x^{12} + 64x^{10}$

Chapter 6 Answers

Problem 1. $-5(9x + 4) = -5(9x) - 5(4) = -45x - 20$

Problem 2. $-x(4x^2 - 2) = -x(4x^2) - x(-2) = -4x^3 + 2x$

Problem 3. $-4x^2(3x^3 + 9x) = -4x^2(3x^3) - 4x^2(9x) = -12x^5 - 36x^3$

Problem 4. $-(-x - 6) = -1(-x) - 1(-6) = x + 6$

Problem 5. $-6x(-8x + 1) = -6x(-8x) - 6x(1) = 48x^2 - 6x$

Problem 6. $-5x^3(7x^2 + 6x) = -5x^3(7x^2) - 5x^3(6x) = -35x^5 - 30x^4$

Problem 7. $-x(x - 2) = -x(x) - x(-2) = -x^2 + 2x$

Problem 8. $-3x(-3x^3 - 4x^2) = -3x(-3x^3) - 3x(-4x^2) = 9x^4 + 12x^3$

Problem 9. $-x^2(6x^2 + 5x) = -x^2(6x^2) - x^2(5x) = -6x^4 - 5x^3$

Problem 10. $-6(4x^2 - 8x) = -6(4x^2) - 6(-8x) = -24x^2 + 48x$

Problem 11. $-9x(7x^3 + 8x) = -9x(7x^3) - 9x(8x) = -63x^4 - 72x^2$

Problem 12. $-2x^3(-4x^3 - 12) = -2x^3(-4x^3) - 2x^3(-12) = 8x^6 + 24x^3$

Problem 13. $-6x^2(-5x^4 + 4x^2) = -6x^2(-5x^4) - 6x^2(4x^2) = 30x^6 - 24x^4$

Problem 14. $-8x^4(6x^2 + 5x) = -8x^4(6x^2) - 8x^4(5x) = -48x^6 - 40x^5$

Problem 15. $-7x^2(3x - 5) = -7x^2(3x) - 7x^2(-5) = -21x^3 + 35x^2$

Problem 16. $-10x^5(-9x^4 - 10x^3) = -10x^5(-9x^4) - 10x^5(-10x^3) = 90x^9 + 100x^8$

Problem 17. $-3(x^2 + 5x + 9) = -3(x^2) - 3(5x) - 3(9) = -3x^2 - 15x - 27$

Problem 18. $-(2x^4 - 7x^2 + 9) = -1(2x^4) - 1(-7x^2) - 1(9) = -2x^4 + 7x^2 - 9$

Problem 19. $-8x^2(-x^3 - 7x^2 - 6x) = -8x^2(-x^3) - 8x^2(-7x^2) - 8x^2(-6x) =$
$8x^5 + 56x^4 + 48x^3$

Problem 20. $-6x(-9x^2 + 8x - 5) = -6x(-9x^2) - 6x(8x) - 6x(-5) = 54x^3 - 48x^2 + 30x$

Problem 21. $-5x^3(6x^5 + 2x^3 + 8x) = -5x^3(6x^5) - 5x^3(2x^3) - 5x^3(8x) =$
$-30x^8 - 10x^6 - 40x^4$

Problem 22. $-x(-4x^6 - 5x^5 - x) = -x(-4x^6) - x(-5x^5) - x(-x) = 4x^7 + 5x^6 + x^2$

Problem 23. $-9(-2x^2 - 6x + 7) = -9(-2x^2) - 9(-6x) - 9(7) = 18x^2 + 54x - 63$

Problem 24. $-2x^2(-7x^4 - 8x^3 + 4x^2) = -2x^2(-7x^4) - 2x^2(-8x^3) - 2x^2(4x^2) =$
$14x^6 + 16x^5 - 8x^4$

Problem 25. $-4x^5(6x^5 - 6x^4 + 4x) = -4x^5(6x^5) - 4x^5(-6x^4) - 4x^5(4x) =$
$-24x^{10} + 24x^9 - 16x^6$

Problem 26. $-9x^2(-4x^7 - 3x^2 - 5x) = -9x^2(-4x^7) - 9x^2(-3x^2) - 9x^2(-5x) =$
$36x^9 + 27x^4 + 45x^3$

Problem 27. $-6x(2x^3 + x^2 + 4x) = -6x(2x^3) - 6x(x^2) - 6x(4x) = -12x^4 - 6x^3 - 24x^2$

Problem 28. $-7x^2(-3x^3 + 2x - 6) = -7x^2(-3x^3) - 7x^2(2x) - 7x^2(-6) =$
$21x^5 - 14x^3 + 42x^2$

Problem 29. $-8x(x^3 - 4x^2 + 7x) = -8x(x^3) - 8x(-4x^2) - 8x(7x) =$
$-8x^4 + 32x^3 - 56x^2$

Problem 30. $-9x^5(-6x^6 + 5x^3 - 8x) = -9x^5(-6x^6) - 9x^5(5x^3) - 9x^5(-8x) =$
$54x^{11} - 45x^8 + 72x^6$

Problem 31. $-2x^3(-9x^7 + 3x^5 - 5x^3 + x) =$
$-2x^3(-9x^7) - 2x^3(3x^5) - 2x^3(-5x^3) - 2x^3(x) = 18x^{10} - 6x^8 + 10x^6 - 2x^4$

Problem 32. $-3x^2(-5x^4 + 6x^3 - 7x^2 - 2x) =$
$-3x^2(-5x^4) - 3x^2(6x^3) - 3x^2(-7x^2) - 3x^2(-2x) = 15x^6 - 18x^5 + 21x^4 + 6x^3$

Problem 33. $-5x^4(8x^3 + 6x^2 + x - 5) = -5x^4(8x^3) - 5x^4(6x^2) - 5x^4(x) - 5x^4(-5) =$
$-40x^7 - 30x^6 - 5x^5 + 25x^4$

Problem 34. $-9x^9(-11x^8 + 10x^7 + 15x^5 - 9x^2) =$
$-9x^9(-11x^8) - 9x^9(10x^7) - 9x^9(15x^5) - 9x^9(-9x^2) = 99x^{17} - 90x^{16} - 135x^{14} + 81x^{11}$

Chapter 7 Answers

Problem 1. $\frac{2}{3}(12x + 18) = \frac{2}{3}(12x) + \frac{2}{3}(18) = 8x + 12$

Problem 2. $\frac{1}{4}(36x^2 - 28x) = \frac{1}{4}(36x^2) + \frac{1}{4}(-28x) = 9x^2 - 7x$

Problem 3. $-\frac{3x}{2}(3x + 5) = -\frac{3x}{2}(3x) - \frac{3x}{2}(5) = -\frac{9}{2}x^2 - \frac{15}{2}x$

Problem 4. $\frac{x}{3}(-x^2 + 6x) = \frac{x}{3}(-x^2) + \frac{x}{3}(6x) = -\frac{x^3}{3} + 2x^2$

Problem 5. $-\frac{4x^2}{5}\left(\frac{10}{3}x - \frac{5}{2}\right) = -\frac{4x^2}{5}\left(\frac{10}{3}x\right) - \frac{4x^2}{5}\left(-\frac{5}{2}\right) = -\frac{8}{3}x^3 + 2x^2$

Notes: $\frac{40}{15} = \frac{40\div5}{15\div5} = \frac{8}{3}$ and $\frac{20}{10} = 2$.

Problem 6. $\frac{5}{4}\left(-\frac{6}{5}x^3 - \frac{18}{7}x\right) = \frac{5}{4}\left(-\frac{6}{5}x^3\right) + \frac{5}{4}\left(-\frac{18}{7}x\right) = -\frac{3}{2}x^3 - \frac{45}{14}x$

Notes: $\frac{30}{20} = \frac{3}{2}$ and $\frac{90}{28} = \frac{45}{14}$.

Problem 7. $\frac{5x}{2}\left(\frac{9}{10}x^2 - \frac{6}{25}x\right) = \frac{5x}{2}\left(\frac{9}{10}x^2\right) + \frac{5x}{2}\left(-\frac{6}{25}x\right) = \frac{9}{4}x^3 - \frac{3}{5}x^2$

Notes: $\frac{45}{20} = \frac{45\div5}{20\div5} = \frac{9}{4}$ and $\frac{30}{50} = \frac{3}{5}$.

Problem 8. $-\frac{x^3}{6}(9x^4 - 3x^2 + 15) = -\frac{x^3}{6}(9x^4) - \frac{x^3}{6}(-3x^2) - \frac{x^3}{6}(15) = -\frac{3}{2}x^7 + \frac{x^5}{2} - \frac{5}{2}x^3$

Notes: $\frac{9}{6} = \frac{9\div3}{6\div3} = \frac{3}{2}$ and $\frac{15}{6} = \frac{15\div3}{6\div3} = \frac{5}{2}$.

Problem 9. $\frac{9x}{4}\left(-\frac{x^2}{3} + \frac{5x}{6} - \frac{10}{27}\right) = \frac{9x}{4}\left(-\frac{x^2}{3}\right) + \frac{9x}{4}\left(\frac{5x}{6}\right) + \frac{9x}{4}\left(-\frac{10}{27}\right) = -\frac{3}{4}x^3 + \frac{15}{8}x^2 - \frac{5}{6}x$

Notes: $\frac{9}{12} = \frac{9\div3}{12\div3} = \frac{3}{4}, \frac{45}{24} = \frac{45\div3}{24\div3} = \frac{15}{8}$, and $\frac{90}{108} = \frac{90\div18}{108\div18} = \frac{5}{6}$. Tip: $\frac{9}{4}\frac{10}{27} = \frac{9}{27}\frac{10}{4} = \frac{1}{3}\frac{5}{2} = \frac{5}{6}$.

Problem 10. $\frac{3(4x-6)}{2} = \frac{3}{2}(4x) + \frac{3}{2}(-6) = 6x - 9$

Problem 11. $-4\frac{9x^2+2}{3} = -\frac{4}{3}(9x^2) - \frac{4}{3}(2) = -12x^2 - \frac{8}{3}$

Problem 12. $\frac{2x(-3x+4)}{5} = \frac{2x}{5}(-3x) + \frac{2x}{5}(4) = -\frac{6}{5}x^2 + \frac{8}{5}x$

Problem 13. $\frac{9x^3-6x}{4} = \frac{1}{4}(9x^3) + \frac{1}{4}(-6x) = \frac{9}{4}x^3 - \frac{3}{2}x$

Problem 14. $\frac{4x^2+2x-8}{2} = \frac{1}{2}(4x^2) + \frac{1}{2}(2x) + \frac{1}{2}(-8) = 2x^2 + x - 4$

Problem 15. $-\frac{x(x-6)}{3} = -\frac{x}{3}(x) - \frac{x}{3}(-6) = -\frac{x^2}{3} + 2x$

Problem 16. $x\frac{9x^4+2x^2-12x}{3} = \frac{x}{3}(9x^4) + \frac{x}{3}(2x^2) + \frac{x}{3}(-12x) = 3x^5 + \frac{2}{3}x^3 - 4x^2$

Problem 17. $\frac{3x^2(6x^5-3x^3+x)}{2} = \frac{3x^2}{2}(6x^5) + \frac{3x^2}{2}(-3x^3) + \frac{3x^2}{2}(x) = 9x^7 - \frac{9}{2}x^5 + \frac{3}{2}x^3$

Problem 18. $-9\frac{6x^2-10}{4} = -\frac{9}{4}(6x^2) - \frac{9}{4}(-10) = -\frac{27}{2}x^2 + \frac{45}{2}$

Notes: $\frac{54}{4} = \frac{54 \div 2}{4 \div 2} = \frac{27}{2}$ and $\frac{90}{4} = \frac{90 \div 2}{4 \div 2} = \frac{45}{2}$.

Problem 19. $5x\frac{2x^2-x+3}{9} = \frac{5x}{9}(2x^2) + \frac{5x}{9}(-x) + \frac{5x}{9}(3) = \frac{10}{9}x^3 - \frac{5}{9}x^2 + \frac{5x}{3}$

Chapter 8 Answers

Problem 1. $\frac{x^8}{x^2} = x^{8-2} = x^6$

Problem 2. $\frac{x^9}{x^4} = x^{9-4} = x^5$

Problem 3. $\frac{x^6}{x} = \frac{x^6}{x^1} = x^{6-1} = x^5$

Problem 4. $\frac{x^7}{x^9} = x^{7-9} = x^{-2} = \frac{1}{x^2}$

Problem 5. $\frac{x^6}{x^3} = x^{6-3} = x^3$

Problem 6. $\frac{x^7}{x^6} = x^{7-6} = x^1 = x$

Problem 7. $\frac{x}{x^2} = \frac{x^1}{x^2} = x^{1-2} = x^{-1} = \frac{1}{x}$

Problem 8. $\frac{x^3}{x^3} = x^{3-3} = x^0 = 1$

Problem 9. $\frac{x^7 x^5}{x^4} = \frac{x^{7+5}}{x^4} = \frac{x^{12}}{x^4} = x^{12-4} = x^8$

Problem 10. $\frac{x^2 x^2}{x^9} = \frac{x^{2+2}}{x^9} = \frac{x^4}{x^9} = x^{4-9} = x^{-5} = \frac{1}{x^5}$

Problem 11. $\frac{x^8 x}{x^5} = \frac{x^8 x^1}{x^5} = \frac{x^{8+1}}{x^5} = \frac{x^9}{x^5} = x^{9-5} = x^4$

Problem 12. $\frac{x^9}{x^6 x^2} = \frac{x^9}{x^{6+2}} = \frac{x^9}{x^8} = x^{9-8} = x^1 = x$

Problem 13. $\frac{x^{12}}{x^4 x^2} = \frac{x^{12}}{x^{4+2}} = \frac{x^{12}}{x^6} = x^{12-6} = x^6$

Problem 14. $\frac{x^3 x}{x^4} = \frac{x^3 x^1}{x^4} = \frac{x^{3+1}}{x^4} = \frac{x^4}{x^4} = x^{4-4} = x^0 = 1$

Problem 15. $\frac{x^2}{x^9 x} = \frac{x^2}{x^9 x^1} = \frac{x^2}{x^{9+1}} = \frac{x^2}{x^{10}} = x^{2-10} = x^{-8} = \frac{1}{x^8}$

Problem 16. $\frac{x^7 x^4}{x^8 x^2} = \frac{x^{7+4}}{x^{8+2}} = \frac{x^{11}}{x^{10}} = x^{11-10} = x^1 = x$

Problem 17. $\frac{x^9 x^8}{x^7 x^6} = \frac{x^{9+8}}{x^{7+6}} = \frac{x^{17}}{x^{13}} = x^{17-13} = x^4$

Problem 18. $\frac{x^4 x}{x^3 x^3} = \frac{x^4 x^1}{x^3 x^3} = \frac{x^{4+1}}{x^{3+3}} = \frac{x^5}{x^6} = x^{5-6} = x^{-1} = \frac{1}{x}$

Problem 19. $\frac{x^6 x^3}{x^8 x^7} = \frac{x^{6+3}}{x^{8+7}} = \frac{x^9}{x^{15}} = x^{9-15} = x^{-6} = \frac{1}{x^6}$

Problem 20. $\frac{x^6 x^5 x^4}{x^3} = \frac{x^{6+5+4}}{x^3} = \frac{x^{15}}{x^3} = x^{15-3} = x^{12}$

Problem 21. $\frac{x^8}{x^9 x^7 x^5} = \frac{x^8}{x^{9+7+5}} = \frac{x^8}{x^{21}} = x^{8-21} = x^{-13} = \frac{1}{x^{13}}$

Problem 22. $\dfrac{x^{10}}{x^3 x^2 x} = \dfrac{x^{10}}{x^3 x^2 x^1} = \dfrac{x^{10}}{x^{3+2+1}} = \dfrac{x^{10}}{x^6} = x^{10-6} = x^4$

Problem 23. $\dfrac{x^9 x^6 x^3}{x} = \dfrac{x^{9+6+3}}{x^1} = \dfrac{x^{18}}{x^1} = x^{18-1} = x^{17}$

Problem 24. $\dfrac{x^8 x^6 x^2}{x^5 x^3} = \dfrac{x^{8+6+2}}{x^{5+3}} = \dfrac{x^{16}}{x^8} = x^{16-8} = x^8$

Problem 25. $\dfrac{x^9 x^8}{x^5 x^3 x} = \dfrac{x^9 x^8}{x^5 x^3 x^1} = \dfrac{x^{9+8}}{x^{5+3+1}} = \dfrac{x^{17}}{x^9} = x^{17-9} = x^8$

Problem 26. $\dfrac{x^9 x^9 x^9}{x^7 x^7 x^7} = \dfrac{x^{9+9+9}}{x^{7+7+7}} = \dfrac{x^{27}}{x^{21}} = x^{27-21} = x^6$

Problem 27. $\dfrac{x^6 x^4 x^2}{x^7 x^5 x^5} = \dfrac{x^{6+4+2}}{x^{7+5+5}} = \dfrac{x^{12}}{x^{17}} = x^{12-17} = x^{-5} = \dfrac{1}{x^5}$

Problem 28. $\dfrac{x^8 x^4 x}{x^9 x^7 x^6} = \dfrac{x^8 x^4 x^1}{x^9 x^7 x^6} = \dfrac{x^{8+4+1}}{x^{9+7+6}} = \dfrac{x^{13}}{x^{22}} = x^{13-22} = x^{-9} = \dfrac{1}{x^9}$

Chapter 9 Answers

Problem 1. $4x^5 + 6x^2 = 2x^2(2x^3 + 3)$

Problem 2. $16x^2 + 40x = 8x(2x + 5)$

Problem 3. $12x - 36 = 12(x - 3)$

Problem 4. $8x^4 + 4x^2 = 4x^2(2x^2 + 1)$

Problem 5. $14x^5 - 19x^3 = x^3(14x^2 - 19)$

Problem 6. $7x^2 + 42x = 7x(x + 6)$

Problem 7. $25x^6 - 15x^4 = 5x^4(5x^2 - 3)$

Problem 8. $14x^6 + 21x^2 = 7x^2(2x^4 + 3)$

Problem 9. $21x - 28 = 7(3x - 4)$

Problem 10. $3x^3 + 27x = 3x(x^2 + 9)$

Problem 11. $18x^9 - 24x^5 = 6x^5(3x^4 - 4)$

Problem 12. $45x^4 + 15x^3 = 15x^3(3x + 1)$

Problem 13. $16x^6 + 20x = 4x(4x^5 + 5)$

Problem 14. $8x^6 + 14x^4 = 2x^4(4x^2 + 7)$

Problem 15. $48x^4 - 40 = 8(6x^4 - 5)$

Problem 16. $2x^2 + x = x(2x + 1)$

Problem 17. $54x^8 - 42x^3 = 6x^3(9x^5 - 7)$

Problem 18. $18x^7 + 15x = 3x(6x^6 + 5)$

Problem 19. $24x^7 - 36x^3 = 12x^3(2x^4 - 3)$

Problem 20. $40x^5 + 32x^2 = 8x^2(5x^3 + 4)$

Problem 21. $18x^8 - 40x^4 = 2x^4(9x^4 - 20)$

Problem 22. $48x^9 - 80x^6 = 16x^6(3x^3 - 5)$

Chapter 10 Answers

Problem 1. $12x^5 + 18x^3 + 4x = 2x(6x^4 + 9x^2 + 2)$

Problem 2. $18x^9 + 36x^8 - 24x^7 = 6x^7(3x^2 + 6x - 4)$

Problem 3. $6x^2 - 9x + 12 = 3(2x^2 - 3x + 4)$

Problem 4. $7x^6 - 3x^4 - 2x^2 = x^2(7x^4 - 3x^2 - 2)$

Problem 5. $11x^5 + 22x^4 + 33x^3 = 11x^3(x^2 + 2x + 3)$

Problem 6. $14x^7 + 28x^4 - 42x = 14x(x^6 + 2x^3 - 3)$

Problem 7. $12x^4 - 24x + 8 = 4(3x^4 - 6x + 2)$

Problem 8. $40x^8 - 80x^5 - 48x^2 = 8x^2(5x^6 - 10x^3 - 6)$

Problem 9. $35x^{11} + 25x^8 - 10x^5 = 5x^5(7x^6 + 5x^3 - 2)$

Problem 10. $27x^5 + 81x^4 + 9x^3 = 9x^3(3x^2 + 9x + 1)$

Problem 11. $48x^7 - 36x^5 + 48x^3 = 12x^3(4x^4 - 3x^2 + 4)$

Problem 12. $24x^8 + 48x^6 - 72x^4 = 24x^4(x^4 + 2x^2 - 3)$

Problem 13. $6x^3 - 8x^2 - 9x = x(6x^2 - 8x - 9)$

Problem 14. $72x^5 - 54x^3 + 108x^2 = 18x^2(4x^3 - 3x + 6)$

Problem 15. $60x^7 - 105x^6 - 75x^5 = 15x^5(4x^2 - 7x - 5)$

Problem 16. $180x^4 + 240x^2 + 270 = 30(6x^4 + 8x^2 + 9)$

Problem 17. $105x^{10} + 63x^7 - 168x^3 = 21x^3(5x^7 + 3x^4 - 8)$

Problem 18. $16x^7 - 32x^6 + 80x^5 = 16x^5(x^2 - 2x + 5)$

Problem 19. $56x^5 - 72x^3 - 32x = 8x(7x^4 - 9x^2 - 4)$

Problem 20. $180x^{14} + 108x^9 - 288x^4 = 36x^4(5x^{10} + 3x^5 - 8)$

Problem 21. $160x^{15} - 80x^{13} - 140x^9 = 20x^9(8x^6 - 4x^4 - 7)$

Problem 22. $88x^8 + 22x^7 + 66x^6 = 22x^6(4x^2 + x + 3)$

Problem 23. $54x^4 - 72x^2 + 45 = 9(6x^4 - 8x^2 + 5)$

Problem 24. $28x^{11} + 42x^8 - 56x^5 = 14x^5(2x^6 + 3x^3 - 4)$

Problem 25. $125x^4 - 225x^3 - 175x^2 = 25x^2(5x^2 - 9x - 7)$

Problem 26. $36x^{12} + 12x^{10} - 18x^7 = 6x^7(6x^5 + 2x^3 - 3)$

Problem 27. $18x^7 + 72x^5 + 144x^3 = 18x^3(x^4 + 4x^2 + 8)$

Problem 28. $120x^{10} - 200x^9 + 360x^8 = 40x^8(3x^2 - 5x + 9)$

Problem 29. $256x^4 + 96x^3 - 160x = 32x(8x^3 + 3x^2 - 5)$

Problem 30. $100x^{10} - 80x^7 - 30x^4 = 10x^4(10x^6 - 8x^3 - 3)$

Problem 31. $375x^{10} + 525x^6 + 450x^2 = 75x^2(5x^8 + 7x^4 + 6)$

Problem 32. $56x^9 + 16x^8 - 56x^7 = 8x^7(7x^2 + 2x - 7)$

Problem 33. $216x^{13} - 216x^8 + 54x^3 = 54x^3(4x^{10} - 4x^5 + 1)$

Problem 34. $288x^{14} - 384x^{11} - 240x^8 = 48x^8(6x^6 - 8x^3 - 5)$

Chapter 11 Answers

Problem 1. $24x^5 + 18x^4 + 6x^3 + 3x^2 = 3x^2(8x^3 + 6x^2 + 2x + 1)$

Problem 2. $24x^7 + 42x^5 - 30x^3 + 54x = 6x(4x^6 + 7x^4 - 5x^2 + 9)$

Problem 3. $30x^3 - 5x^2 + 40x - 25 = 5(6x^3 - x^2 + 8x - 5)$

Problem 4. $50x^8 - 40x^7 - 34x^6 - 6x^5 = 2x^5(25x^3 - 20x^2 - 17x - 3)$

Problem 5. $x^7 + x^4 + x^2 + x = x(x^6 + x^3 + x + 1)$

Problem 6. $24x^{12} + 72x^9 + 36x^6 - 24x^3 = 12x^3(2x^9 + 6x^6 + 3x^3 - 2)$

Problem 7. $96x^8 - 80x^6 + 40x^4 + 64x^2 = 8x^2(12x^6 - 10x^4 + 5x^2 + 8)$

Problem 8. $28x^4 - 36x^3 + 12x^2 - 24x = 4x(7x^3 - 9x^2 + 3x - 6)$

Problem 9. $45x^{10} - 72x^8 - 18x^6 + 36x^4 = 9x^4(5x^6 - 8x^4 - 2x^2 + 4)$

Problem 10. $48x^5 + 96x^4 - 16x^3 + 80x^2 = 16x^2(3x^3 + 6x^2 - x + 5)$

Problem 11. $50x^{14} - 150x^{11} - 75x^8 - 175x^5 = 25x^5(2x^9 - 6x^6 - 3x^3 - 7)$

Problem 12. $49x^9 - 70x^7 + 35x^5 + 56x^3 = 7x^3(7x^6 - 10x^4 + 5x^2 + 8)$

Problem 13. $80x^{10} + 100x^9 - 60x^8 - 120x^7 = 20x^7(4x^3 + 5x^2 - 3x - 6)$

Problem 14. $112x^{22} - 28x^{18} - 70x^{14} + 84x^{10} = 14x^{10}(8x^{12} - 2x^8 - 5x^4 + 6)$

Problem 15. $72x^{11} + 120x^{10} - 192x^9 - 48x^8 = 24x^8(3x^3 + 5x^2 - 8x - 2)$

Problem 16. $192x^7 - 256x^5 - 256x^3 - 96x = 32x(6x^6 - 8x^4 - 8x^2 - 3)$

Problem 17. $108x^{24} + 243x^{19} - 162x^{14} - 189x^9 = 27x^9(4x^{15} + 9x^{10} - 6x^5 - 7)$

Chapter 12 Answers

Problem 1. $-8x^3 - 12x^2 = -4x^2(2x + 3)$

Problem 2. $-2x + 8 = -2(x - 4)$

Problem 3. $-30x^5 - 18x^3 = -6x^3(5x^2 + 3)$

Problem 4. $-x^8 - x^4 = -x^4(x^4 + 1)$

Problem 5. $-6x^9 + 21x^6 = -3x^6(2x^3 - 7)$

Problem 6. $-36x^2 + 27x = -9x(4x - 3)$

Problem 7. $-9x^6 - 5x^5 = -x^5(9x + 5)$

Problem 8. $-15x^5 + 40x^3 = -5x^3(3x^2 - 8)$

Problem 9. $-49x^9 - 63x^3 = -7x^3(7x^6 + 9)$

Problem 10. $-48x^4 - 40 = -8(6x^4 + 5)$

Problem 11. $-36x^8 - 81x^5 - 36x^2 = -9x^2(4x^6 + 9x^3 + 4)$

Problem 12. $-24x^2 + 48x + 32 = -8(3x^2 - 6x - 4)$

Problem 13. $-5x^9 + 25x^7 - 15x^5 = -5x^5(x^4 - 5x^2 + 3)$

Problem 14. $x^{10} - 3x^7 - 2x^4 = -x^4(-x^6 + 3x^3 + 2)$

Problem 15. $-30x^7 - 46x^5 - 22x^3 = -2x^3(15x^4 + 23x^2 + 11)$

Problem 16. $-64x^4 - 32x^3 + 48x = -16x(4x^3 + 2x^2 - 3)$

Problem 17. $-49x^8 + 35x^6 - 56x^4 = -7x^4(7x^4 - 5x^2 + 8)$

Problem 18. $-18x^4 + 54x^3 + 12x^2 = -6x^2(3x^2 - 9x - 2)$

Problem 19. $4x^8 - 6x^4 - 9 = -1(-4x^8 + 6x^4 + 9) = -(-4x^8 + 6x^4 + 9)$

Problem 20. $-72x^5 + 24x^3 - 60x = -12x(6x^4 - 2x^2 + 5)$

Problem 21. $-144x^5 - 162x^4 + 72x^3 = -18x^3(8x^2 + 9x - 4)$

Problem 22. $-105x^{15} - 63x^{10} - 147x^5 = -21x^5(5x^{10} + 3x^5 + 7)$

Problem 23. $-8x^6 + 16x^5 - 12x^4 + 18x^3 = -2x^3(4x^3 - 8x^2 + 6x - 9)$

Problem 24. $-6x^{12} + 7x^{10} + 2x^8 + x^6 = -x^6(6x^6 - 7x^4 - 2x^2 - 1)$

Problem 25. $-35x^3 - 21x^2 - 7x - 28 = -7(5x^3 + 3x^2 + x + 4)$

Problem 26. $36x^{12} - 24x^8 - 20x^6 - 8x^4 = -4x^4(-9x^8 + 6x^4 + 5x^2 + 2)$

Problem 27. $-48x^{10} + 54x^7 + 42x^4 - 30x = -6x(8x^9 - 9x^6 - 7x^3 + 5)$

Problem 28. $-27x^5 + 36x^4 - 81x^3 + 72x^2 = -9x^2(3x^3 - 4x^2 + 9x - 8)$

Problem 29. $35x^7 - 10x^5 - 20x^4 - 35x = -5x(-7x^6 + 2x^4 + 4x^3 + 7)$

Problem 30. $-6x^{11} + 24x^9 + 18x^7 + 9x^5 = -3x^5(2x^6 - 8x^4 - 6x^2 - 3)$

Problem 31. $-8x^5 + 56x^4 - 64x^3 + 48x^2 = -8x^2(x^3 - 7x^2 + 8x - 6)$

Problem 32. $88x^{16} - 55x^{13} - 33x^{10} - 77x^7 = -11x^7(-8x^9 + 5x^6 + 3x^3 + 7)$

Problem 33. $-126x^5 + 56x^4 - 70x^3 - 56x^2 = -14x^2(9x^3 - 4x^2 + 5x + 4)$

Problem 34. $57x^{16} - 152x^{12} - 114x^8 - 171x^4 = -19x^4(-3x^{12} + 8x^8 + 6x^4 + 9)$

Chapter 13 Answers

Problem 1. $(x+3)(x+2) = x(x) + x(2) + 3(x) + 3(2) = x^2 + 2x + 3x + 6 = x^2 + 5x + 6$

Problem 2. $(x+6)(x+4) = x(x) + x(4) + 6(x) + 6(4) = x^2 + 4x + 6x + 24 =$
$x^2 + 10x + 24$

Problem 3. $(x-8)(x-1) = x(x) + x(-1) - 8(x) - 8(-1) = x^2 - x - 8x + 8 =$
$x^2 - 9x + 8$

Problem 4. $(x+3)(x-9) = x(x) + x(-9) + 3(x) + 3(-9) = x^2 - 9x + 3x - 27 =$
$x^2 - 6x - 27$

Problem 5. $(x+9)(x-3) = x(x) + x(-3) + 9(x) + 9(-3) = x^2 - 3x + 9x - 27 =$
$x^2 + 6x - 27$

Problem 6. $(x+4)(x+7) = x(x) + x(7) + 4(x) + 4(7) = x^2 + 7x + 4x + 28 =$
$x^2 + 11x + 28$

Problem 7. $(x-9)(x+6) = x(x) + x(6) - 9(x) - 9(6) = x^2 + 6x - 9x - 54 =$
$x^2 - 3x - 54$

Problem 8. $(x+7)(x-5) = x(x) + x(-5) + 7(x) + 7(-5) = x^2 - 5x + 7x - 35 =$
$x^2 + 2x - 35$

Problem 9. $(x+8)(x+8) = x(x) + x(8) + 8(x) + 8(8) = x^2 + 8x + 8x + 64 =$
$x^2 + 16x + 64$

Problem 10. $(x-5)(x-3) = x(x) + x(-3) - 5(x) - 5(-3) = x^2 - 3x - 5x + 15 =$
$x^2 - 8x + 15$

Problem 11. $(x-4)(x+8) = x(x) + x(8) - 4(x) - 4(8) = x^2 + 8x - 4x - 32 =$
$x^2 + 4x - 32$

Problem 12. $(-x+8)(x+5) = -x(x) - x(5) + 8(x) + 8(5) = -x^2 - 5x + 8x + 40 =$
$-x^2 + 3x + 40$

Problem 13. $(6-x)(8-x) = 6(8) + 6(-x) - x(8) - x(-x) = 48 - 6x - 8x + x^2 =$
$x^2 - 14x + 48$

Problem 14. $(x+1)(x+9) = x(x) + x(9) + 1(x) + 1(9) = x^2 + 9x + x + 9 =$
$x^2 + 10x + 9$

Problem 15. $(x-4)(x+4) = x(x) + x(4) - 4(x) - 4(4) = x^2 + 4x - 4x - 16 =$
$x^2 - 16$

Problem 16. $(-x+4)(x+2) = -x(x) - x(2) + 4(x) + 4(2) = -x^2 - 2x + 4x + 8 =$
$-x^2 + 2x + 8$

Problem 17. $(x-5)(-x-5) = x(-x) + x(-5) - 5(-x) - 5(-5) = -x^2 - 5x + 5x + 25 = -x^2 + 25$

Problem 18. $(x+6)(x+3) = x(x) + x(3) + 6(x) + 6(3) = x^2 + 3x + 6x + 18 = x^2 + 9x + 18$

Problem 19. $(9-x)(2-x) = 9(2) + 9(-x) - x(2) - x(-x) = 18 - 9x - 2x + x^2 = x^2 - 11x + 18$

Problem 20. $(-x-9)(-x-9) = -x(-x) - x(-9) - 9(-x) - 9(-9) = x^2 + 9x + 9x + 81 = x^2 + 18x + 81$

Problem 21. $(x-3)(x+4) = x(x) + x(4) - 3(x) - 3(4) = x^2 + 4x - 3x - 12 = x^2 + x - 12$

Problem 22. $(x+6)(x-3) = x(x) + x(-3) + 6(x) + 6(-3) = x^2 - 3x + 6x - 18 = x^2 + 3x - 18$

Problem 23. $(x-5)(x-2) = x(x) + x(-2) - 5(x) - 5(-2) = x^2 - 2x - 5x + 10 = x^2 - 7x + 10$

Problem 24. $(x+9)(x+7) = x(x) + x(7) + 9(x) + 9(7) = x^2 + 7x + 9x + 63 = x^2 + 16x + 63$

Problem 25. $(-x+7)(x+6) = -x(x) - x(6) + 7(x) + 7(6) = -x^2 - 6x + 7x + 42 = -x^2 + x + 42$

Problem 26. $(-x-3)(x+9) = -x(x) - x(9) - 3(x) - 3(9) = -x^2 - 9x - 3x - 27 = -x^2 - 12x - 27$

Problem 27. $(x+7)(x-9) = x(x) + x(-9) + 7(x) + 7(-9) = x^2 - 9x + 7x - 63 = x^2 - 2x - 63$

Problem 28. $(x+6)(x+5) = x(x) + x(5) + 6(x) + 6(5) = x^2 + 5x + 6x + 30 = x^2 + 11x + 30$

Problem 29. $(-x+5)(x+8) = -x(x) - x(8) + 5(x) + 5(8) = -x^2 - 8x + 5x + 40 = -x^2 - 3x + 40$

Problem 30. $(-x-1)(-x-1) = -x(-x) - x(-1) - 1(-x) - 1(-1) = x^2 + x + x + 1 = x^2 + 2x + 1$

Problem 31. $(x-5)(-x-9) = x(-x) + x(-9) - 5(-x) - 5(-9) = -x^2 - 9x + 5x + 45 = -x^2 - 4x + 45$

Problem 32. $(x+4)(x+2) = x(x) + x(2) + 4(x) + 4(2) = x^2 + 2x + 4x + 8 = x^2 + 6x + 8$

Problem 33. $(2-x)(7-x) = 2(7) + 2(-x) - x(7) - x(-x) = 14 - 2x - 7x + x^2 = x^2 - 9x + 14$

Problem 34. $(-x-8)(-x-9) = -x(-x) - x(-9) - 8(-x) - 8(-9) =$
$x^2 + 9x + 8x + 72 = x^2 + 17x + 72$

Problem 35. $(6x+5)(3x+4) = 6x(3x) + 6x(4) + 5(3x) + 5(4) =$
$18x^2 + 24x + 15x + 20 = 18x^2 + 39x + 20$

Problem 36. $(2x+3)(7x+9) = 2x(7x) + 2x(9) + 3(7x) + 3(9) =$
$14x^2 + 18x + 21x + 27 = 14x^2 + 39x + 27$

Problem 37. $(7x+4)(5x-9) = 7x(5x) + 7x(-9) + 4(5x) + 4(-9) =$
$35x^2 - 63x + 20x - 36 = 35x^2 - 43x - 36$

Problem 38. $(3x-5)(3x-6) = 3x(3x) + 3x(-6) - 5(3x) - 5(-6) =$
$9x^2 - 18x - 15x + 30 = 9x^2 - 33x + 30$

Problem 39. $(4x-7)(6x+8) = 4x(6x) + 4x(8) - 7(6x) - 7(8) =$
$24x^2 + 32x - 42x - 56 = 24x^2 - 10x - 56$

Problem 40. $(5x+2)(6x+4) = 5x(6x) + 5x(4) + 2(6x) + 2(4) =$
$30x^2 + 20x + 12x + 8 = 30x^2 + 32x + 8$

Problem 41. $(6x+7)(2x-9) = 6x(2x) + 6x(-9) + 7(2x) + 7(-9) =$
$12x^2 - 54x + 14x - 63 = 12x^2 - 40x - 63$

Problem 42. $(8x+9)(8x+9) = 8x(8x) + 8x(9) + 9(8x) + 9(9) =$
$64x^2 + 72x + 72x + 81 = 64x^2 + 144x + 81$

Problem 43. $(2x-6)(6x-2) = 2x(6x) + 2x(-2) - 6(6x) - 6(-2) =$
$12x^2 - 4x - 36x + 12 = 12x^2 - 40x + 12$

Problem 44. $(9x+5)(3x-7) = 9x(3x) + 9x(-7) + 5(3x) + 5(-7) =$
$27x^2 - 63x + 15x - 35 = 27x^2 - 48x - 35$

Problem 45. $(7x+6)(8x+9) = 7x(8x) + 7x(9) + 6(8x) + 6(9) =$
$56x^2 + 63x + 48x + 54 = 56x^2 + 111x + 54$

Problem 46. $(-5x+3)(4x-8) = -5x(4x) - 5x(-8) + 3(4x) + 3(-8) =$
$-20x^2 + 40x + 12x - 24 = -20x^2 + 52x - 24$

Problem 47. $(4+7x)(x-8) = 4(x) + 4(-8) + 7x(x) + 7x(-8) = 4x - 32 + 7x^2 - 56x =$
$7x^2 - 52x - 32$

Problem 48. $(6x+3)(6-9x) = 6x(6) + 6x(-9x) + 3(6) + 3(-9x) =$
$36x - 54x^2 + 18 - 27x = -54x^2 + 9x + 18$

Problem 49. $(8+2x)(5x+7) = 8(5x) + 8(7) + 2x(5x) + 2x(7) =$
$40x + 56 + 10x^2 + 14x = 10x^2 + 54x + 56$

Problem 50. $(3x+5)(3x-5) = 3x(3x) + 3x(-5) + 5(3x) + 5(-5) =$
$9x^2 - 15x + 15x - 25 = 9x^2 - 25$

Problem 51. $(8 + 4x)(7 - 4x) = 8(7) + 8(-4x) + 4x(7) + 4x(-4x) =$
$56 - 32x + 28x - 16x^2 = -16x^2 - 4x + 56$

Problem 52. $(-9x - 8)(-6x + 1) = -9x(-6x) - 9x(1) - 8(-6x) - 8(1) =$
$54x^2 - 9x + 48x - 8 = 54x^2 + 39x - 8$

Problem 53. $(4 + 5x)(2x - 3) = 4(2x) + 4(-3) + 5x(2x) + 5x(-3) =$
$8x - 12 + 10x^2 - 15x = 10x^2 - 7x - 12$

Problem 54. $(8x - 3)(5 - 4x) = 8x(5) + 8x(-4x) - 3(5) - 3(-4x) =$
$40x - 32x^2 - 15 + 12x = -32x^2 + 52x - 15$

Problem 55. $(-2x + 9)(3x - 7) = -2x(3x) - 2x(-7) + 9(3x) + 9(-7) =$
$-6x^2 + 14x + 27x - 63 = -6x^2 + 41x - 63$

Problem 56. $(6 + 7x)(5 - 2x) = 6(5) + 6(-2x) + 7x(5) + 7x(-2x) =$
$30 - 12x + 35x - 14x^2 = -14x^2 + 23x + 30$

Problem 57. $(x^2 + x)(x + 1) = x^2(x) + x^2(1) + x(x) + x(1) = x^3 + x^2 + x^2 + x =$
$x^3 + 2x^2 + x$

Problem 58. $(4x^2 - 5)(3x + 6) = 4x^2(3x) + 4x^2(6) - 5(3x) - 5(6) =$
$12x^3 + 24x^2 - 15x - 30$

Problem 59. $(9x^2 - 8)(6x + 7) = 9x^2(6x) + 9x^2(7) - 8(6x) - 8(7) =$
$54x^3 + 63x^2 - 48x - 56$

Problem 60. $(3x^2 + 5)(4x^2 - 6) = 3x^2(4x^2) + 3x^2(-6) + 5(4x^2) + 5(-6) =$
$12x^4 - 18x^2 + 20x^2 - 30 = 12x^4 + 2x^2 - 30$

Problem 61. $(7x^2 - 6x)(4x - 2) = 7x^2(4x) + 7x^2(-2) - 6x(4x) - 6x(-2) =$
$28x^3 - 14x^2 - 24x^2 + 12x = 28x^3 - 38x^2 + 12x$

Problem 62. $(2x^2 + 2x)(2x + 2) = 2x^2(2x) + 2x^2(2) + 2x(2x) + 2x(2) =$
$4x^3 + 4x^2 + 4x^2 + 4x = 4x^3 + 8x^2 + 4x$

Problem 63. $(7x^2 + 9)(4x^3 - 6) = 7x^2(4x^3) + 7x^2(-6) + 9(4x^3) + 9(-6) =$
$28x^5 - 42x^2 + 36x^3 - 54 = 28x^5 + 36x^3 - 42x^2 - 54$

Problem 64. $(x^2 + 1)(x^2 - 1) = x^2(x^2) + x^2(-1) + 1(x^2) + 1(-1) = x^4 - x^2 + x^2 - 1 =$
$x^4 - 1$

Problem 65. $(6x + 5)(7x^2 + 3) = 6x(7x^2) + 6x(3) + 5(7x^2) + 5(3) =$
$42x^3 + 18x + 35x^2 + 15 = 42x^3 + 35x^2 + 18x + 15$

Problem 66. $(2x^2 - 6x)(2x - 3) = 2x^2(2x) + 2x^2(-3) - 6x(2x) - 6x(-3) =$
$4x^3 - 6x^2 - 12x^2 + 18x = 4x^3 - 18x^2 + 18x$

Problem 67. $(6x^2 + 7)(8x^2 - 9) = 6x^2(8x^2) + 6x^2(-9) + 7(8x^2) + 7(-9) =$
$48x^4 - 54x^2 + 56x^2 - 63 = 48x^4 + 2x^2 - 63$

Problem 68. $(4x^2 - 2)(3x + 5) = 4x^2(3x) + 4x^2(5) - 2(3x) - 2(5) =$
$12x^3 + 20x^2 - 6x - 10$

Problem 69. $(5x^3 + 3x)(4x^2 - 3x) = 5x^3(4x^2) + 5x^3(-3x) + 3x(4x^2) + 3x(-3x) =$
$20x^5 - 15x^4 + 12x^3 - 9x^2$

Problem 70. $(4x^2 + 4x)(6x^2 + 6) = 4x^2(6x^2) + 4x^2(6) + 4x(6x^2) + 4x(6) =$
$24x^4 + 24x^2 + 24x^3 + 24x = 24x^4 + 24x^3 + 24x^2 + 24x$

Problem 71. $(5x^3 - 9x)(3x - 5) = 5x^3(3x) + 5x^3(-5) - 9x(3x) - 9x(-5) =$
$15x^4 - 25x^3 - 27x^2 + 45x$

Problem 72. $(7x^2 + 5x)(6x^3 - 5x) = 7x^2(6x^3) + 7x^2(-5x) + 5x(6x^3) + 5x(-5x) =$
$42x^5 - 35x^3 + 30x^4 - 25x^2 = 42x^5 + 30x^4 - 35x^3 - 25x^2$

Problem 73. $(4x^3 - 7x)(6x^2 + 2) = 4x^3(6x^2) + 4x^3(2) - 7x(6x^2) - 7x(2) =$
$24x^5 + 8x^3 - 42x^3 - 14x = 24x^5 - 34x^3 - 14x$

Problem 74. $(2x^2 + 4)(6x^3 - 5) = 2x^2(6x^3) + 2x^2(-5) + 4(6x^3) + 4(-5) =$
$12x^5 - 10x^2 + 24x^3 - 20 = 12x^5 + 24x^3 - 10x^2 - 20$

Problem 75. $(9x^3 - 5x)(8x^4 - 4x^2) = 9x^3(8x^4) + 9x^3(-4x^2) - 5x(8x^4) - 5x(-4x^2) =$
$72x^7 - 36x^5 - 40x^5 + 20x^3 = 72x^7 - 76x^5 + 20x^3$

Problem 76. $(6x^3 + 3x)(4x^2 + 3x) = 6x^3(4x^2) + 6x^3(3x) + 3x(4x^2) + 3x(3x) =$
$24x^5 + 18x^4 + 12x^3 + 9x^2$

Problem 77. $(-6x + 7)(-5x^3 - 8) = -6x(-5x^3) - 6x(-8) + 7(-5x^3) + 7(-8) =$
$30x^4 + 48x - 35x^3 - 56 = 30x^4 - 35x^3 + 48x - 56$

Problem 78. $(5x^2 + 3)(4x^3 - 5) = 5x^2(4x^3) + 5x^2(-5) + 3(4x^3) + 3(-5) =$
$20x^5 - 25x^2 + 12x^3 - 15 = 20x^5 + 12x^3 - 25x^2 - 15$

Problem 79. $(2x^3 - 6x)(3x^4 + 8x^2) = 2x^3(3x^4) + 2x^3(8x^2) - 6x(3x^4) - 6x(8x^2) =$
$6x^7 + 16x^5 - 18x^5 - 48x^3 = 6x^7 - 2x^5 - 48x^3$

Problem 80. $(4x^3 + 3x)(4x^4 - 3x^2) = 4x^3(4x^4) + 4x^3(-3x^2) + 3x(4x^4) + 3x(-3x^2) =$
$16x^7 - 12x^5 + 12x^5 - 9x^3 = 16x^7 - 9x^3$

Problem 81. $(9x^4 - 7x^2)(8x^3 - 6x) = 9x^4(8x^3) + 9x^4(-6x) - 7x^2(8x^3) - 7x^2(-6x) =$
$72x^7 - 54x^5 - 56x^5 + 42x^3 = 72x^7 - 110x^5 + 42x^3$

Problem 82. $(6x^3 - 9x^2)(7 + 9x^2) = 6x^3(7) + 6x^3(9x^2) - 9x^2(7) - 9x^2(9x^2) =$
$42x^3 + 54x^5 - 63x^2 - 81x^4 = 54x^5 - 81x^4 + 42x^3 - 63x^2$

Chapter 14 Answers

Problem 1. $(x + 1)^2 = x^2 + 2(x)(1) + 1^2 = x^2 + 2x + 1$

Problem 2. $(x + 6)^2 = x^2 + 2(x)(6) + 6^2 = x^2 + 12x + 36$

Problem 3. $(x - 2)^2 = x^2 + 2(x)(-2) + (-2)^2 = x^2 - 4x + 4$

Problem 4. $(x - 3)^2 = x^2 + 2(x)(-3) + (-3)^2 = x^2 - 6x + 9$

Problem 5. $(-x + 7)^2 = (-x)^2 + 2(-x)(7) + 7^2 = x^2 - 14x + 49$

Problem 6. $(x + 9)^2 = x^2 + 2(x)(9) + 9^2 = x^2 + 18x + 81$

Problem 7. $(8 - x)^2 = 8^2 + 2(8)(-x) + (-x)^2 = 64 - 16x + x^2 = x^2 - 16x + 64$

Problem 8. $(x - 12)^2 = x^2 + 2(x)(-12) + (-12)^2 = x^2 - 24x + 144$

Problem 9. $(20 + x)^2 = 20^2 + 2(20)(x) + x^2 = 400 + 40x + x^2 = x^2 + 40x + 400$

Problem 10. $(-x - 5)^2 = (-x)^2 + 2(-x)(-5) + (-5)^2 = x^2 + 10x + 25$

Problem 11. $(3x + 6)^2 = (3x)^2 + 2(3x)(6) + 6^2 = 9x^2 + 36x + 36$

Problem 12. $(6x - 4)^2 = (6x)^2 + 2(6x)(-4) + (-4)^2 = 36x^2 - 48x + 16$

Problem 13. $(7x - 6)^2 = (7x)^2 + 2(7x)(-6) + (-6)^2 = 49x^2 - 84x + 36$

Problem 14. $(4x + 5)^2 = (4x)^2 + 2(4x)(5) + 5^2 = 16x^2 + 40x + 25$

Problem 15. $(8x - 5)^2 = (8x)^2 + 2(8x)(-5) + (-5)^2 = 64x^2 - 80x + 25$

Problem 16. $(2x + 7)^2 = (2x)^2 + 2(2x)(7) + 7^2 = 4x^2 + 28x + 49$

Problem 17. $(9x - 9)^2 = (9x)^2 + 2(9x)(-9) + (-9)^2 = 81x^2 - 162x + 81$

Problem 18. $(-5x + 8)^2 = (-5x)^2 + 2(-5x)(8) + 8^2 = 25x^2 - 80x + 64$

Problem 19. $(7 + 3x)^2 = 7^2 + 2(7)(3x) + (3x)^2 = 49 + 42x + 9x^2 = 9x^2 + 42x + 49$

Problem 20. $(-6x - 4)^2 = (-6x)^2 + 2(-6x)(-4) + (-4)^2 = 36x^2 + 48x + 16$

Problem 21. $(1 - 8x)^2 = 1^2 + 2(1)(-8x) + (-8x)^2 = 1 - 16x + 64x^2 = 64x^2 - 16x + 1$

Problem 22. $(-9x + 2)^2 = (-9x)^2 + 2(-9x)(2) + 2^2 = 81x^2 - 36x + 4$

Problem 23. $(3x^2 + 4)^2 = (3x^2)^2 + 2(3x^2)(4) + 4^2 = 9x^4 + 24x^2 + 16$

Note: Recall the rule $(ax^m)^n = a^n x^{mn}$ from the second part of Chapter 2.

Problem 24. $(8x^2 - 5x)^2 = (8x^2)^2 + 2(8x^2)(-5x) + (-5x)^2 = 64x^4 - 80x^3 + 25x^2$

Problem 25. $(-6x^3 + 1)^2 = (-6x^3)^2 + 2(-6x^3)(1) + 1^2 = 36x^6 - 12x^3 + 1$

Problem 26. $(9x^4 + 7x^2)^2 = (9x^4)^2 + 2(9x^4)(7x^2) + (7x^2)^2 = 81x^8 + 126x^6 + 49x^4$

Problem 27. $(5x^2 - 5)^2 = (5x^2)^2 + 2(5x^2)(-5) + (-5)^2 = 25x^4 - 50x^2 + 25$

Problem 28. $(-2x^3 - 8x^2)^2 = (-2x^3)^2 + 2(-2x^3)(-8x^2) + (-8x^2)^2 =$
$4x^6 + 32x^5 + 64x^4$

Problem 29. $(4 + 3x^2)^2 = 4^2 + 2(4)(3x^2) + (3x^2)^2 = 16 + 24x^2 + 9x^4 = 9x^4 + 24x^2 + 16$

Problem 30. $(7x^3 - 6x)^2 = (7x^3)^2 + 2(7x^3)(-6x) + (-6x)^2 = 49x^6 - 84x^4 + 36x^2$

Problem 31. $(1 - x^5)^2 = 1^2 + 2(1)(-x^5) + (-x^5)^2 = 1 - 2x^5 + x^{10} = x^{10} - 2x^5 + 1$

Problem 32. $(-8x^4 + 4x^2)^2 = (-8x^4)^2 + 2(-8x^4)(4x^2) + (4x^2)^2 = 64x^8 - 64x^6 + 16x^4$

Problem 33. $(x - x^3)^2 = x^2 + 2(x)(-x^3) + (-x^3)^2 = x^2 - 2x^4 + x^6 = x^6 - 2x^4 + x^2$

Problem 34. $(9x^5 - 8x^4)^2 = (9x^5)^2 + 2(9x^5)(-8x^4) + (-8x^4)^2 = 81x^{10} - 144x^9 + 64x^8$

Problem 35. $x^2 + 12x + 36 = x^2 + 2x(6) + 6^2 = (x + 6)^2$

Problem 36. $x^2 - 8x + 16 = x^2 + 2x(-4) + (-4)^2 = (x - 4)^2$

Alternate answers: $(4 - x)^2$, $(-x + 4)^2$, and $(-4 + x)^2$ are equivalent to $(x - 4)^2$.

Problem 37. $x^2 + 18x + 81 = x^2 + 2x(9) + 9^2 = (x + 9)^2$

Problem 38. $9x^2 + 24x + 16 = (3x)^2 + 2(3x)(4) + 4^2 = (3x + 4)^2$

Problem 39. $16x^2 - 72x + 81 = (4x)^2 + 2(4x)(-9) + (-9)^2 = (4x - 9)^2$

Alternate answers: $(9 - 4x)^2$, $(-4x + 9)^2$, and $(-9 + 4x)^2$ are equivalent to $(4x - 9)^2$.

Problem 40. $64x^2 - 80x + 25 = (8x)^2 + 2(8x)(-5) + (-5)^2 = (8x - 5)^2$

Alternate answers: $(5 - 8x)^2$, $(-8x + 5)^2$, and $(-5 + 8x)^2$ are equivalent to $(8x - 5)^2$.

Problem 41. $25x^2 + 70x + 49 = (5x)^2 + 2(5x)(7) + 7^2 = (5x + 7)^2$

Problem 42. $49x^2 + 70x + 25 = (7x)^2 + 2(7x)(5) + 5^2 = (7x + 5)^2$

Problem 43. $x^4 - 2x^2 + 1 = (x^2)^2 + 2(x^2)(-1) + (-1)^2 = (x^2 - 1)^2$

Alternate answers: $(1 - x^2)^2$, $(-x^2 + 1)^2$, and $(-1 + x^2)^2$ are equivalent to $(x^2 - 1)^2$.

Problem 44. $9x^4 + 36x^3 + 36x^2 = (3x^2)^2 + 2(3x^2)(6x) + (6x)^2 = (3x^2 + 6x)^2$

Alternate answers: $9x^2(x + 2)^2$ and $9(x^2 + 2x)^2$ are equivalent to $(3x^2 + 6x)^2$.

Chapter 15 Answers

Problem 1. $(x + 2)(x - 2) = x^2 - 2^2 = x^2 - 4$

Problem 2. $(x + 5)(x - 5) = x^2 - 5^2 = x^2 - 25$

Problem 3. $(x + 4)(x - 4) = x^2 - 4^2 = x^2 - 16$

Problem 4. $(4 + x)(4 - x) = 4^2 - x^2 = 16 - x^2 = -x^2 + 16$

Problem 5. $(x - 7)(x + 7) = (x + 7)(x - 7) = x^2 - 7^2 = x^2 - 49$

Note: $ab = ba$ because multiplication is commutative (meaning that the order in which numbers are multiplied does not matter; for example, $3 \times 4 = 4 \times 3 = 12$).

Problem 6. $(6 + x)(6 - x) = 6^2 - x^2 = 36 - x^2 = -x^2 + 36$

Problem 7. $(x - 5)(x + 5) = (x + 5)(x - 5) = x^2 - 5^2 = x^2 - 25$

Problem 8. $(9 + x)(9 - x) = 9^2 - x^2 = 81 - x^2 = -x^2 + 81$

Problem 9. $(-x + 1)(-x - 1) = (-x)^2 - 1^2 = x^2 - 1$

Problem 10. $(8 - x)(8 + x) = (8 + x)(8 - x) = 8^2 - x^2 = 64 - x^2 = -x^2 + 64$

Problem 11. $(5x + 2)(5x - 2) = (5x)^2 - 2^2 = 25x^2 - 4$

Problem 12. $(9x + 7)(9x - 7) = (9x)^2 - 7^2 = 81x^2 - 49$

Problem 13. $(3x - 4)(3x + 4) = (3x + 4)(3x - 4) = (3x)^2 - 4^2 = 9x^2 - 16$

Problem 14. $(6x + 3)(6x - 3) = (6x)^2 - 3^2 = 36x^2 - 9$

Problem 15. $(8x - 6)(8x + 6) = (8x + 6)(8x - 6) = (8x)^2 - 6^2 = 64x^2 - 36$

Problem 16. $(4x + 4)(4x - 4) = (4x)^2 - 4^2 = 16x^2 - 16$

Problem 17. $(7 + 2x)(7 - 2x) = 7^2 - (2x)^2 = 49 - 4x^2 = -4x^2 + 49$

Problem 18. $(9x + 5)(9x - 5) = (9x)^2 - 5^2 = 81x^2 - 25$

Problem 19. $(-6x + 3)(6x + 3) = (3 - 6x)(3 + 6x) = (3 + 6x)(3 - 6x) = 3^2 - (6x)^2 = 9 - 36x^2 = -36x^2 + 9$ Note: $-6x + 3 = 3 - 6x$.

Problem 20. $(5 - 4x)(5 + 4x) = (5 + 4x)(5 - 4x) = 5^2 - (4x)^2 = 25 - 16x^2 = -16x^2 + 25$

Problem 21. $(-8x - 9)(-8x + 9) = (-8x)^2 - 9^2 = 64x^2 - 81$

Problem 22. $(7x + 6)(7x - 6) = (7x)^2 - 6^2 = 49x^2 - 36$

Problem 23. $(x^2 + 1)(x^2 - 1) = (x^2)^2 - 1^2 = x^4 - 1$

Problem 24. $(x^2 + 7)(x^2 - 7) = (x^2)^2 - 7^2 = x^4 - 49$

Problem 25. $(x^3 - 2)(x^3 + 2) = (x^3 + 2)(x^3 - 2) = (x^3)^2 - 2^2 = x^6 - 4$

Problem 26. $(x^3 - 4)(x^3 + 4) = (x^3 + 4)(x^3 - 4) = (x^3)^2 - 4^2 = x^6 - 16$

Problem 27. $(5x^3 + 3)(5x^3 - 3) = (5x^3)^2 - 3^2 = 25x^6 - 9$

Problem 28. $(10x^4 + 10)(10x^4 - 10) = (10x^4)^2 - 10^2 = 100x^8 - 100$

Problem 29. $(7x^4 - 7)(7x^4 + 7) = (7x^4 + 7)(7x^4 - 7) = (7x^4)^2 - 7^2 = 49x^8 - 49$

Problem 30. $(2x^5 + 5x^3)(2x^5 - 5x^3) = (2x^5)^2 - (5x^3)^2 = 4x^{10} - 25x^6$

Problem 31. $(4x^6 + 6x^4)(4x^6 - 6x^4) = (4x^6)^2 - (6x^4)^2 = 16x^{12} - 36x^8$

Problem 32. $(9 - 6x^6)(9 + 6x^6) = (9 + 6x^6)(9 - 6x^6) = 9^2 - (6x^6)^2 = 81 - 36x^{12} = -36x^{12} + 81$

Problem 33. $(6x^7 + 2)(6x^7 - 2) = (6x^7)^2 - 2^2 = 36x^{14} - 4$

Problem 34. $(-10 + 8x^8)(-10 - 8x^8) = (-10)^2 - (8x^8)^2 = 100 - 64x^{16} = -64x^{16} + 100$

Problem 35. $x^2 - 4 = x^2 - 2^2 = (x + 2)(x - 2)$

Problem 36. $x^2 - 25 = x^2 - 5^2 = (x + 5)(x - 5)$

Problem 37. $25 - x^2 = 5^2 - x^2 = (5 + x)(5 - x)$

Problem 38. $64 - x^2 = 8^2 - x^2 = (8 + x)(8 - x)$

Problem 39. $x^2 - 36 = x^2 - 6^2 = (x + 6)(x - 6)$

Problem 40. $49x^2 - 81 = (7x)^2 - 9^2 = (7x + 9)(7x - 9)$

Problem 41. $16x^2 - 121 = (4x)^2 - 11^2 = (4x + 11)(4x - 11)$

Problem 42. $144 - 81x^2 = 12^2 - (9x)^2 = (12 + 9x)(12 - 9x)$

Problem 43. $x^4 - 1 = (x^2)^2 - 1^2 = (x^2 + 1)(x^2 - 1) = (x^2 + 1)(x + 1)(x - 1)$

Note: $x^2 - 1 = x^2 - 1^2 = (x + 1)(x - 1)$.

Problem 44. $9x^2 - 64 = (3x)^2 - 8^2 = (3x + 8)(3x - 8)$

Chapter 16 Answers

Problem 1. $x = 6$ Check: $5x - 12 = 5(6) - 12 = 30 - 12 = 18$

Problem 2. $x = 8$ Check: $x + 9 = 8 + 9 = 17$

Problem 3. $x = 2$ Check: $8 + 3x = 8 + 3(2) = 8 + 6 = 14$ agrees with $7x = 7(2) = 14$

Problem 4. $x = 9$ Check: $6x = 6(9) = 54$

Problem 5. $x = 3$ Check: $9 - x = 9 - 3 = 6$ agrees with $2x = 2(3) = 6$

Problem 6. $x = 4$ Check: $4x = 4(4) = 16$ agrees with $28 - 3x = 28 - 3(4) = 28 - 12 = 16$

Problem 7. $x = 1$ Check: $8x + 6 = 8(1) + 6 = 8 + 6 = 14$

Problem 8. $x = 7$ Check: $21 + 2x = 21 + 2(7) = 21 + 14 = 35$ agrees with $5x = 5(7) = 35$

Problem 9. $x = 7$ Check: $7x - 28 = 7(7) - 28 = 49 - 28 = 21$ agrees with $3x = 3(7) = 21$

Problem 10. $x = -8$ Check: $4 - x = 4 - (-8) = 4 + 8 = 12$

Problem 11. $x = 5$ Check: $6x + 18 = 6(5) + 18 = 30 + 18 = 48$

Problem 12. $x = 9$ Check: $12x - 72 = 12(9) - 72 = 108 - 72 = 36$ agrees with $4x = 4(9) = 36$

Problem 13. $x = -2$ Check: $6 + 5x = 6 + 5(-2) = 6 - 10 = -4$ agrees with $2x = 2(-2) = -4$

Problem 14. $x = 1$ Check: $4x = 4(1) = 4$ agrees with $8 - 4x = 8 - 4(1) = 8 - 4 = 4$

Problem 15. $x = -8$ Check: $16x = 16(-8) = -128$ agrees with $9x - 56 = 9(-8) - 56 = -72 - 56 = -128$

Problem 16. $x = -3$ Check: $6 - 8x = 6 - 8(-3) = 6 + 24 = 30$

Problem 17. $x = 3$ Check: $15 - x = 15 - 3 = 12$ agrees with $4x = 4(3) = 12$

Problem 18. $x = -10$ Check: $14x + 90 = 14(-10) + 90 = -140 + 90 = -50$ agrees with $5x = 5(-10) = -50$

Problem 19. $x = 7$ Check: $5x + 5 = 5(7) + 5 = 35 + 5 = 40$ agrees with $3x + 19 = 3(7) + 19 = 21 + 19 = 40$

Problem 20. $x = 4$ Check: $9x - 8 = 9(4) - 8 = 36 - 8 = 28$ agrees with $4x + 12 = 4(4) + 12 = 16 + 12 = 28$

Problem 21. $x = 8$ Check: $6x - 32 = 6(8) - 32 = 48 - 32 = 16$ agrees with $32 - 2x = 32 - 2(8) = 32 - 16 = 16$

Problem 22. $x = 2$ Check: $16 + 7x = 16 + 7(2) = 16 + 14 = 30$ agrees with $36 - 3x = 36 - 3(2) = 36 - 6 = 30$

Problem 23. $x = 2$ Check: $3 - 7x = 3 - 7(2) = 3 - 14 = -11$ agrees with $7 - 9x = 7 - 9(2) = 7 - 18 = -11$

Problem 24. $x = -1$ Check: $3 - x = 3 - (-1) = 3 + 1 = 4$ agrees with $5 + x = 5 + (-1) = 5 - 1 = 4$

Problem 25. $x = -9$ Check: $14 + 4x = 14 + 4(-9) = 14 - 36 = -22$ agrees with $8x + 50 = 8(-9) + 50 = -72 + 50 = -22$

Problem 26. $x = 8$ Check: $2x + 28 = 2(8) + 28 = 16 + 28 = 44$ agrees with $8x - 20 = 8(8) - 20 = 64 - 20 = 44$

Problem 27. $x = 7$ Check: $-2x + 18 = -2(7) + 18 = -14 + 18 = 4$ agrees with $60 - 8x = 60 - 8(7) = 60 - 56 = 4$

Problem 28. $x = 9$ Check: $13x - 31 = 13(9) - 31 = 117 - 31 = 86$ agrees with $6x + 32 = 6(9) + 32 = 54 + 32 = 86$

Problem 29. $x = 0$ Check: $2x - 21 = 2(0) - 21 = -21$ agrees with $5x - 21 = 5(0) - 21 = -21$

Problem 30. $x = -5$ Check: $6 + 4x = 6 + 4(-5) = 6 - 20 = -14$ agrees with $31 + 9x = 31 + 9(-5) = 31 - 45 = -14$

Problem 31. $x = -4$ Check: $44 - 2x = 44 - 2(-4) = 44 + 8 = 52$ agrees with $12 - 10x = 12 - 10(-4) = 12 + 40 = 52$

Problem 32. $x = 3$ Check: $x + 12 = 3 + 12 = 15$ agrees with $9x - 12 = 9(3) - 12 = 27 - 12 = 15$

Problem 33. $x = 2$ Check: $-3 + 9x = -3 + 9(2) = -3 + 18 = 15$ agrees with $3x + 9 = 3(2) + 9 = 6 + 9 = 15$

Problem 34. $x = 1$ Check: $7x - 11 = 7(1) - 11 = 7 - 11 = -4$ agrees with $1 - 5x = 1 - 5(1) = 1 - 5 = -4$

Problem 35. $x = 12$ Check: $4 + 20x = 4 + 20(12) = 4 + 240 = 244$ agrees with $100 + 12x = 100 + 12(12) = 100 + 144 = 244$

Problem 36. $x = -9$ Check: $32 - 8x = 32 - 8(-9) = 32 + 72 = 104$ agrees with $41 - 7x = 41 - 7(-9) = 41 + 63 = 104$

Problem 37. $x = -13$ Check: $9x - 63 = 9(-13) - 63 = -117 - 63 = -180$ agrees with $16x + 28 = 16(-13) + 28 = -208 + 28 = -180$

Problem 38. $x = -20$ Check: $22x + 150 - 16x = 22(-20) + 150 - 16(-20) = -440 + 150 + 320 = 30$

Problem 39. $x = -11$ Check: $-1 - x = -1 - (-11) = -1 + 11 = 10$ agrees with $10x + 120 = 10(-11) + 120 = -110 + 120 = 10$

Problem 40. $x = -6$ Check: $-11x + 80 = -11(-6) + 80 = 66 + 80 = 146$ agrees with $8 - 23(-6) = 8 + 138 = 146$

Problem 41. $x = -9$ Check: $4x - 27 = 4(-9) - 27 = -36 - 27 = -63$ agrees with $-2x + 9x = -2(-9) + 9(-9) = 18 - 81 = -63$

Problem 42. $x = 10$ Check: $100 - 7x = 100 - 7(10) = 100 - 70 = 30$ agrees with $8x - 50 = 8(10) - 50 = 80 - 50 = 30$

Problem 43. $x = 3$ Check: $6x - 4 + 2x - 7 + x = 6(3) - 4 + 2(3) - 7 + 3 =$
$18 - 4 + 6 - 7 + 3 = 16$ agrees with $8 + 9x - 9 - 3x - 1 = 8 + 9(3) - 9 - 3(3) - 1 =$
$8 + 27 - 9 - 9 - 1 = 16$

Problem 44. $x = -8$ Check: $14 - 12x - 35 + 8x - 28 = 14 - 12(-8) - 35 + 8(-8) - 28 =$
$14 + 96 - 35 - 64 - 28 = -17$ agrees with $5x - 63 + 9x + 70 - 11x =$
$5(-8) - 63 + 9(-8) + 70 - 11(-8) = -40 - 63 - 72 + 70 + 88 = -17$

Problem 45. $x = 2$ Check: $-24x - 96 + 36x + 144 - 48x =$
$-24(2) - 96 + 36(2) + 144 - 48(2) = -48 - 96 + 72 + 144 - 96 = -24$ agrees with
$60x - 108 - 72x - 60 + 84x = 60(2) - 108 - 72(2) - 60 + 84(2) =$
$120 - 108 - 144 - 60 + 168 = -24$

Chapter 17 Answers

Problem 1. $\frac{x}{4} + \frac{2}{3} - \frac{x}{6} + \frac{1}{2} = \left(\frac{3x}{12} - \frac{2x}{12}\right) + \left(\frac{4}{6} + \frac{3}{6}\right) = \frac{x}{12} + \frac{7}{6}$

Problem 2. $\frac{2x^2}{3} + \frac{5}{6} + \frac{4x^2}{9} - \frac{1}{3} = \left(\frac{6x^2}{9} + \frac{4x^2}{9}\right) + \left(\frac{5}{6} - \frac{2}{6}\right) = \frac{10x^2}{9} + \frac{3}{6} = \frac{10x^2}{9} + \frac{1}{2}$

Problem 3. $x^2 + 2x - \frac{x^2}{5} - \frac{x}{4} = \left(\frac{5x^2}{5} - \frac{x^2}{5}\right) + \left(\frac{8x}{4} - \frac{x}{4}\right) = \frac{4x^2}{5} + \frac{7x}{4}$

Problem 4. $\frac{7x}{4} - \frac{3}{2} + \frac{9x}{8} - \frac{2}{3} = \left(\frac{14x}{8} + \frac{9x}{8}\right) + \left(-\frac{9}{6} - \frac{4}{6}\right) = \frac{23x}{8} - \frac{13}{6}$

Problem 5. $\frac{5x^2}{6} - \frac{1}{4} + \frac{2x^2}{9} + \frac{5}{2} = \left(\frac{15x^2}{18} + \frac{4x^2}{18}\right) + \left(-\frac{1}{4} + \frac{10}{4}\right) = \frac{19x^2}{18} + \frac{9}{4}$

Problem 6. $\frac{x^3}{2} - \frac{x^3}{4} + \frac{x^2}{2} - \frac{x^2}{6} = \left(\frac{2x^3}{4} - \frac{x^3}{4}\right) + \left(\frac{3x^2}{6} - \frac{x^2}{6}\right) = \frac{x^3}{4} + \frac{2x^2}{6} = \frac{x^3}{4} + \frac{x^2}{3}$

Note: The terms were ordered differently in this problem. The first two terms both have x^3 and the last two terms both have x^2.

Problem 7. $-\frac{2x}{7} - \frac{7}{6} + \frac{5x}{2} + \frac{1}{6} = \left(-\frac{4x}{14} + \frac{35x}{14}\right) + \left(-\frac{7}{6} + \frac{1}{6}\right) = \frac{31x}{14} - \frac{6}{6} = \frac{31x}{14} - 1$

Problem 8. $\frac{11x^2}{12} + 3x - \frac{x^2}{4} - \frac{4x}{3} = \left(\frac{11x^2}{12} - \frac{3x^2}{12}\right) + \left(\frac{9x}{3} - \frac{4x}{3}\right) = \frac{8x^2}{12} + \frac{5x}{3} = \frac{2x^2}{3} + \frac{5x}{3}$

Problem 9. $\frac{x^2}{2} + \frac{x^3}{6} - \frac{x^2}{4} + \frac{x^3}{3} = \left(\frac{x^3}{6} + \frac{2x^3}{6}\right) + \left(\frac{2x^2}{4} - \frac{x^2}{4}\right) = \frac{3x^3}{6} + \frac{x^2}{4} = \frac{x^3}{2} + \frac{x^2}{4}$

Problem 10. $\frac{x^2}{8} + 2 + \frac{x^2}{6} - 9 = \left(\frac{3x^2}{24} + \frac{4x^2}{24}\right) + (2 - 9) = \frac{7x^2}{24} - 7$

Problem 11. $\frac{3x}{4} + \frac{x}{6} + \frac{x}{12} = \left(\frac{9x}{12} + \frac{2x}{12} + \frac{x}{12}\right) = \frac{12x}{12} = x$

Problem 12. $7x^2 - \frac{3}{8} - 2x^2 + \frac{5}{8} = (7x^2 - 2x^2) + \left(-\frac{3}{8} + \frac{5}{8}\right) = 5x^2 + \frac{2}{8} = 5x^2 + \frac{1}{4}$

Problem 13. $\frac{x^3}{12} - \frac{x}{18} - \frac{x^3}{6} - \frac{x}{6} = \left(\frac{x^3}{12} - \frac{2x^3}{12}\right) + \left(-\frac{x}{18} - \frac{3x}{18}\right) = -\frac{x^3}{12} - \frac{4x}{18} = -\frac{x^3}{12} - \frac{2x}{9}$

Problem 14. $x^2 + \frac{3x}{8} - \frac{4x^2}{7} - 2x = \left(\frac{7x^2}{7} - \frac{4x^2}{7}\right) + \left(\frac{3x}{8} - \frac{16x}{8}\right) = \frac{3x^2}{7} - \frac{13x}{8}$

Problem 15. $\frac{8x^5}{9} - \frac{7x^2}{8} - \frac{4x^5}{3} + \frac{3x^2}{8} = \left(\frac{8x^5}{9} - \frac{12x^5}{9}\right) + \left(-\frac{7x^2}{8} + \frac{3x^2}{8}\right) = -\frac{4x^5}{9} - \frac{4x^2}{8} = -\frac{4x^5}{9} - \frac{x^2}{2}$

Problem 16. $-\frac{8x}{15} + 6 - \frac{7x}{10} - \frac{9}{2} = \left(-\frac{16x}{30} - \frac{21x}{30}\right) + \left(\frac{12}{2} - \frac{9}{2}\right) = -\frac{37x}{30} + \frac{3}{2}$

Problem 17. $\frac{5}{4} + \frac{2x^2}{5} - \frac{9}{10} - \frac{x^2}{10} = \left(\frac{4x^2}{10} - \frac{x^2}{10}\right) + \left(\frac{25}{20} - \frac{18}{20}\right) = \frac{3x^2}{10} + \frac{7}{20}$

Problem 18. $\frac{6x^4}{11} - \frac{9x}{4} + \frac{11x^4}{6} - \frac{2x}{3} = \left(\frac{36x^4}{66} + \frac{121x^4}{66}\right) + \left(-\frac{27x}{12} - \frac{8x}{12}\right) = \frac{157x^4}{66} - \frac{35x}{12}$

Problem 19. $\frac{7x}{12} - \frac{5x}{24} + \frac{11x}{18} = \left(\frac{42x}{72} - \frac{15x}{72} + \frac{44x}{72}\right) = \frac{71x}{72}$

Problem 20. $\frac{5x^2}{36} - \frac{5}{18} - \frac{7x^2}{24} - \frac{7}{6} = \left(\frac{10x^2}{72} - \frac{21x^2}{72}\right) + \left(-\frac{5}{18} - \frac{21}{18}\right) = -\frac{11x^2}{72} - \frac{26}{18} = -\frac{11x^2}{72} - \frac{13}{9}$

Chapter 18 Answers

Problem 1. $x = \frac{1}{12}$ Check: $x + \frac{2}{3} = \frac{1}{12} + \frac{2}{3} = \frac{1}{12} + \frac{8}{12} = \frac{9}{12} = \frac{3}{4}$

Problem 2. $x = 15$ Check: $\frac{x}{9} = \frac{15}{9} = \frac{15 \div 3}{9 \div 3} = \frac{5}{3}$

Problem 3. $x = \frac{1}{9}$ Check: $\frac{5x}{6} = \frac{5}{6}\left(\frac{1}{9}\right) = \frac{5}{54}$ agrees with $\frac{x}{3} + \frac{1}{18} = \frac{1}{3}\left(\frac{1}{9}\right) + \frac{1}{18} = \frac{1}{27} + \frac{1}{18} = \frac{2}{54} + \frac{3}{54} = \frac{5}{54}$

Problem 4. $x = \frac{7}{8}$ Check: $\frac{3x}{2} - \frac{1}{4} = \frac{3}{2}\left(\frac{7}{8}\right) - \frac{1}{4} = \frac{21}{16} - \frac{4}{16} = \frac{17}{16}$

Problem 5. $x = \frac{3}{16}$ Check: $6x = 6\left(\frac{3}{16}\right) = \frac{18}{16} = \frac{18 \div 2}{16 \div 2} = \frac{9}{8}$

Problem 6. $x = 32$ Check: $\frac{x}{4} = \frac{32}{4} = 8$ agrees with $\frac{x}{6} + \frac{8}{3} = \frac{32}{6} + \frac{8}{3} = \frac{16}{3} + \frac{8}{3} = \frac{24}{3} = 8$

Problem 7. $x = \frac{15}{4}$ Check: $\frac{x}{2} - \frac{1}{4} = \frac{1}{2}\left(\frac{15}{4}\right) - \frac{1}{4} = \frac{15}{8} - \frac{2}{8} = \frac{13}{8}$ agrees with $\frac{x}{3} + \frac{3}{8} = \frac{1}{3}\left(\frac{15}{4}\right) + \frac{3}{8} = \frac{15}{12} + \frac{3}{8} = \frac{30}{24} + \frac{9}{24} = \frac{39}{24} = \frac{39 \div 3}{24 \div 3} = \frac{13}{8}$

Problem 8. $x = \frac{1}{7}$ Check: $\frac{3x}{5} + \frac{8}{15} = \frac{3}{5}\left(\frac{1}{7}\right) + \frac{8}{15} = \frac{3}{35} + \frac{8}{15} = \frac{9}{105} + \frac{56}{105} = \frac{65}{105} = \frac{65 \div 5}{105 \div 5} = \frac{13}{21}$ agrees with $\frac{5x}{6} + \frac{1}{2} = \frac{5}{6}\left(\frac{1}{7}\right) + \frac{1}{2} = \frac{5}{42} + \frac{21}{42} = \frac{26}{42} = \frac{26 \div 2}{42 \div 2} = \frac{13}{21}$

Problem 9. $x = \frac{31}{7}$ Check: $2 - \frac{x}{6} = 2 - \frac{1}{6}\left(\frac{31}{7}\right) = \frac{84}{42} - \frac{31}{42} = \frac{53}{42}$ agrees with $\frac{2x}{9} + \frac{5}{18} = \frac{2}{9}\left(\frac{31}{7}\right) + \frac{5}{18} = \frac{62}{63} + \frac{5}{18} = \frac{124}{126} + \frac{35}{126} = \frac{159}{126} = \frac{159 \div 3}{126 \div 3} = \frac{53}{42}$

Problem 10. $x = \frac{30}{17}$ Check: $4 - \frac{7x}{5} = 4 - \frac{7}{5}\left(\frac{30}{17}\right) = \frac{340}{85} - \frac{210}{85} = \frac{130}{85} = \frac{130 \div 5}{85 \div 5} = \frac{26}{17}$ agrees with $1 + \frac{3x}{10} = 1 + \frac{3}{10}\left(\frac{30}{17}\right) = 1 + \frac{90}{170} = 1 + \frac{9}{17} = \frac{17}{17} + \frac{9}{17} = \frac{26}{17}$

Problem 11. $x = \frac{3}{4}$ Check: $\frac{2x}{3} - \frac{1}{12} = \frac{2}{3}\left(\frac{3}{4}\right) - \frac{1}{12} = \frac{6}{12} - \frac{1}{12} = \frac{5}{12}$ agrees with $x - \frac{1}{3} = \frac{3}{4} - \frac{1}{3} = \frac{9}{12} - \frac{4}{12} = \frac{5}{12}$

Problem 12. $x = \frac{102}{7}$ Check: $-\frac{x}{6} + \frac{5}{2} = -\frac{1}{6}\left(\frac{102}{7}\right) + \frac{5}{2} = -\frac{102}{42} + \frac{105}{42} = \frac{3}{42} = \frac{3 \div 3}{42 \div 3} = \frac{1}{14}$ agrees with $\frac{x}{8} - \frac{7}{4} = \frac{1}{8}\left(\frac{102}{7}\right) - \frac{7}{4} = \frac{102}{56} - \frac{98}{56} = \frac{4}{56} = \frac{4 \div 4}{56 \div 4} = \frac{1}{14}$

Problem 13. $x = \frac{7}{16}$ Check: $\frac{3}{4} - x = \frac{3}{4} - \frac{7}{16} = \frac{12}{16} - \frac{7}{16} = \frac{5}{16}$ agrees with $\frac{1}{6} + \frac{x}{3} = \frac{1}{6} + \frac{1}{3}\left(\frac{7}{16}\right) = \frac{1}{6} + \frac{7}{48} = \frac{8}{48} + \frac{7}{48} = \frac{15}{48} = \frac{15 \div 3}{48 \div 3} = \frac{5}{16}$

Problem 14. $x = \frac{459}{250}$ Check: $\frac{7x}{9} - \frac{3}{8} = \frac{7}{9}\left(\frac{459}{250}\right) - \frac{3}{8} = \frac{3213}{2250} - \frac{3}{8} = \frac{3213 \div 9}{2250 \div 9} - \frac{3}{8} = \frac{357}{250} - \frac{3}{8} = \frac{1428}{1000} - \frac{375}{1000} = \frac{1053}{1000}$ agrees with $\frac{x}{12} + \frac{9}{10} = \frac{1}{12}\left(\frac{459}{250}\right) + \frac{9}{10} = \frac{459}{3000} + \frac{2700}{3000} = \frac{3159}{3000} = \frac{3159 \div 3}{3000 \div 3} = \frac{1053}{1000}$

Problem 15. $x = -\frac{1}{45}$ Check: $\frac{7}{12} + \frac{7x}{2} = \frac{7}{12} + \frac{7}{2}\left(-\frac{1}{45}\right) = \frac{7}{12} - \frac{7}{90} = \frac{105}{180} - \frac{14}{180} = \frac{91}{180}$ agrees with

$\frac{5}{9} + \frac{9x}{4} = \frac{5}{9} + \frac{9}{4}\left(-\frac{1}{45}\right) = \frac{100}{180} - \frac{9}{180} = \frac{91}{180}$

Problem 16. $x = \frac{49}{8}$ Check: $\frac{x}{7} - \frac{7}{3} = \frac{1}{7}\left(\frac{49}{8}\right) - \frac{7}{3} = \frac{7}{8} - \frac{7}{3} = \frac{21}{24} - \frac{56}{24} = -\frac{35}{24}$ agrees with $\frac{7}{6} - \frac{3x}{7} =$

$\frac{7}{6} - \frac{3}{7}\left(\frac{49}{8}\right) = \frac{7}{6} - \frac{21}{8} = \frac{28}{24} - \frac{63}{24} = -\frac{35}{24}$

Notes: $\frac{1}{7}\left(\frac{49}{8}\right) = \frac{49}{7}\frac{1}{8} = \frac{7}{8}$ and $\frac{3}{7}\left(\frac{49}{8}\right) = \frac{49}{7}\frac{3}{8} = \frac{7}{1}\frac{3}{8} = \frac{21}{8}$ because $\frac{49}{7} = 7$.

Problem 17. $x = -\frac{15}{34}$ Check: $\frac{8x}{15} + 3 = \frac{8}{15}\left(-\frac{15}{34}\right) + 3 = -\frac{8}{34} + 3 = -\frac{4}{17} + 3 = -\frac{4}{17} + \frac{51}{17} = \frac{47}{17}$

agrees with $-\frac{7x}{6} + \frac{9}{4} = -\frac{7}{6}\left(-\frac{15}{34}\right) + \frac{9}{4} = \frac{105}{204} + \frac{459}{204} = \frac{564}{204} = \frac{564 \div 12}{204 \div 12} = \frac{47}{17}$

Note: $\frac{8}{15}\left(-\frac{15}{34}\right) = -\frac{8}{15}\frac{15}{34} = -\frac{15}{15}\frac{8}{34} = -\frac{8}{34}$ because the 15's cancel out.

Problem 18. $x = -\frac{17}{6}$ Check: $\frac{4}{9} - \frac{5x}{3} = \frac{4}{9} - \frac{5}{3}\left(-\frac{17}{6}\right) = \frac{8}{18} + \frac{85}{18} = \frac{93}{18} = \frac{93 \div 3}{18 \div 3} = \frac{31}{6}$ agrees with

$\frac{7}{3} - x = \frac{7}{3} - \left(-\frac{17}{6}\right) = \frac{14}{6} + \frac{17}{6} = \frac{31}{6}$

Problem 19. $x = \frac{8}{3}$ Check: $\frac{1}{6} - \frac{x}{8} = \frac{1}{6} - \frac{1}{8}\left(\frac{8}{3}\right) = \frac{4}{24} - \frac{8}{24} = -\frac{4}{24} = -\frac{4 \div 4}{24 \div 4} = -\frac{1}{6}$ agrees with

$\frac{1}{18} - \frac{x}{12} = \frac{1}{18} - \frac{1}{12}\left(\frac{8}{3}\right) = \frac{2}{36} - \frac{8}{36} = -\frac{6}{36} = -\frac{6 \div 6}{36 \div 6} = -\frac{1}{6}$

Problem 20. $x = -\frac{35}{8}$ Check: $\frac{8x}{5} + \frac{9}{4} = \frac{8}{5}\left(-\frac{35}{8}\right) + \frac{9}{4} = -7 + \frac{9}{4} = -\frac{28}{4} + \frac{9}{4} = -\frac{19}{4}$ agrees with

$\frac{2x}{3} - \frac{11}{6} = \frac{2}{3}\left(-\frac{35}{8}\right) - \frac{11}{6} = -\frac{70}{24} - \frac{44}{24} = -\frac{114}{24} = -\frac{114 \div 6}{24 \div 6} = -\frac{19}{4}$

Note: $\frac{8}{5}\left(-\frac{35}{8}\right) = -\frac{8}{5}\frac{35}{8} = -\frac{35}{5}\frac{8}{8} = -7$ because the 8's cancel out and $\frac{35}{5} = 7$.

Problem 21. $x = 2$ Check: $\frac{8x}{9} - \frac{23}{18} = \frac{8}{9}(2) - \frac{23}{18} = \frac{16}{9} - \frac{23}{18} = \frac{32}{18} - \frac{23}{18} = \frac{9}{18} = \frac{9 \div 9}{18 \div 9} = \frac{1}{2}$ agrees

with $\frac{5x}{6} - \frac{7x}{12} = \frac{5}{6}(2) - \frac{7}{12}(2) = \frac{10}{6} - \frac{14}{12} = \frac{20}{12} - \frac{14}{12} = \frac{6}{12} = \frac{6 \div 6}{12 \div 6} = \frac{1}{2}$

Problem 22. $x = \frac{41}{31}$ Check: $-\frac{7}{8} + \frac{3x}{8} = -\frac{7}{8} + \frac{3}{8}\left(\frac{41}{31}\right) = -\frac{217}{248} + \frac{123}{248} = -\frac{94}{248} = -\frac{94 \div 2}{248 \div 2} = -\frac{47}{124}$

agrees with $\frac{5}{6} - \frac{11x}{12} = \frac{5}{6} - \frac{11}{12}\left(\frac{41}{31}\right) = \frac{310}{372} - \frac{451}{372} = -\frac{141}{372} = -\frac{141 \div 3}{372 \div 3} = -\frac{47}{124}$

Chapter 19 Answers

Problem 1. $(3)(2) = 6$ and $3 + 2 = 5$

Problem 2. $(4)(2) = 8$ and $4 + 2 = 6$

Problem 3. $(6)(4) = 24$ and $6 + 4 = 10$

Problem 4. $(5)(-3) = -15$ and $5 + (-3) = 2$

Problem 5. $(-5)(4) = -20$ and $-5 + 4 = -1$

Problem 6. $(5)(-5) = -25$ and $5 + (-5) = 0$

Problem 7. $(-6)(-3) = 18$ and $-6 + (-3) = -9$

Problem 8. $(20)(5) = 100$ and $20 + 5 = 25$

Problem 9. $(12)(4) = 48$ and $12 + 4 = 16$

Problem 10. $(24)(-3) = -72$ and $24 + (-3) = 21$

Problem 11. $(12)(5) = 60$ and $12 + 5 = 17$

Problem 12. $(-20)(3) = -60$ and $-20 + 3 = -17$

Problem 13. $(9)(4) = 36$ and $9 + 4 = 13$

Problem 14. $(-8)(-3) = 24$ and $-8 + (-3) = -11$

Problem 15. $(24)(6) = 144$ and $24 + 6 = 30$

Chapter 20 Answers

Problem 1. $x^2 + 7x + 10 = (x + 5)(x + 2)$

Check: $(x + 5)(x + 2) = x(x) + x(2) + 5(x) + 5(2) = x^2 + 2x + 5x + 10 = x^2 + 7x + 10$

Problem 2. $x^2 - 9x + 8 = (x - 8)(x - 1)$

Check: $(x - 8)(x - 1) = x(x) + x(-1) - 8(x) - 8(-1) = x^2 - x - 8x + 8 = x^2 - 9x + 8$

Problem 3. $x^2 - 4x - 12 = (x - 6)(x + 2)$

Check: $(x - 6)(x + 2) = x(x) + x(2) - 6(x) - 6(2) = x^2 + 2x - 6x - 12 = x^2 - 4x - 12$

Problem 4. $x^2 + 3x - 18 = (x + 6)(x - 3)$

Check: $(x + 6)(x - 3) = x(x) + x(-3) + 6(x) + 6(-3) = x^2 - 3x + 6x - 18 = x^2 + 3x - 18$

Problem 5. $x^2 + 11x + 24 = (x + 8)(x + 3)$

Check: $(x + 8)(x + 3) = x(x) + x(3) + 8(x) + 8(3) = x^2 + 3x + 8x + 24 = x^2 + 11x + 24$

Problem 6. $x^2 - 13x - 30 = (x - 15)(x + 2)$

Check: $(x - 15)(x + 2) = x(x) + x(2) - 15(x) - 15(2) = x^2 + 2x - 15x - 30 = x^2 - 13x - 30$

Problem 7. $x^2 - 17x + 72 = (x - 9)(x - 8)$

Check: $(x - 9)(x - 8) = x(x) + x(-8) - 9(x) - 9(-8) = x^2 - 8x - 9x + 72 = x^2 - 17x + 72$

Problem 8. $x^2 - 20x + 64 = (x - 16)(x - 4)$

Check: $(x - 16)(x - 4) = x(x) + x(-4) - 16(x) - 16(-4) = x^2 - 4x - 16x + 64 = x^2 - 20x + 64$

Problem 9. $x^2 + 3x - 4 = (x + 4)(x - 1)$

Check: $(x + 4)(x - 1) = x(x) + x(-1) + 4(x) + 4(-1) = x^2 - x + 4x - 4 = x^2 + 3x - 4$

Problem 10. $x^2 - 16x + 28 = (x - 14)(x - 2)$

Check: $(x - 14)(x - 2) = x(x) + x(-2) - 14(x) - 14(-2) = x^2 - 2x - 14x + 28 = x^2 - 16x + 28$

Problem 11. $x^2 + 18x + 56 = (x + 14)(x + 4)$

Check: $(x + 14)(x + 4) = x(x) + x(4) + 14(x) + 14(4) = x^2 + 4x + 14x + 56 = x^2 + 18x + 56$

Problem 12. $x^2 - x - 42 = (x - 7)(x + 6)$

Check: $(x - 7)(x + 6) = x(x) + x(6) - 7(x) - 7(6) = x^2 + 6x - 7x - 42 = x^2 - x - 42$

Problem 13. $x^2 - 30x + 81 = (x - 27)(x - 3)$

Check: $(x - 27)(x - 3) = x(x) + x(-3) - 27(x) - 27(-3) = x^2 - 3x - 27x + 81 = x^2 - 30x + 81$

Problem 14. $x^2 + 3x - 40 = (x + 8)(x - 5)$

Check: $(x + 8)(x - 5) = x(x) + x(-5) + 8(x) + 8(-5) = x^2 - 5x + 8x - 40 = x^2 + 3x - 40$

Problem 15. $x^2 - 4x - 45 = (x - 9)(x + 5)$

Check: $(x - 9)(x + 5) = x(x) + x(5) - 9(x) - 9(5) = x^2 + 5x - 9x - 45 = x^2 - 4x - 45$

Problem 16. $x^2 + 12x + 20 = (x + 10)(x + 2)$

Check: $(x + 10)(x + 2) = x(x) + x(2) + 10(x) + 10(2) = x^2 + 2x + 10x + 20 = x^2 + 12x + 20$

Problem 17. $x^2 - 21x + 80 = (x - 16)(x - 5)$

Check: $(x - 16)(x - 5) = x(x) + x(-5) - 16(x) - 16(-5) = x^2 - 5x - 16x + 80 = x^2 - 21x + 80$

Problem 18. $x^2 + 20x - 96 = (x + 24)(x - 4)$

Check: $(x + 24)(x - 4) = x(x) + x(-4) + 24(x) + 24(-4) = x^2 - 4x + 24x - 96 = x^2 + 20x - 96$

WAS THIS BOOK HELPFUL?

Much effort and thought were put into this book, such as:
- Including a variety of useful beginning algebra skills.
- Introducing the main ideas at the beginning of each chapter.
- Providing examples to serve as a helpful guide.
- Tabulating the answers to all of the problems at the back of the book.

If you appreciate the effort that went into making this book possible, there is a simple way that you could show it:

Please take a moment to post an honest review.

For example, you can review this book at Amazon.com or Goodreads.com.

Even a short review can be helpful and will be much appreciated. If you are not sure what to write, following are a few ideas, though it is best to describe what is important to you.
- Was the answer key helpful?
- Were the examples useful?
- Were you able to understand the ideas at the beginning of the chapter?
- Did this book offer good practice for you?
- Would you recommend this book to others? If so, why?

Do you believe that you found a mistake? Please email the author, Chris McMullen, at greekphysics@yahoo.com to ask about it. One of two things will happen:
- You might discover that it wasn't a mistake after all and learn why.
- You might be right, in which case the author will be grateful and future readers will benefit from the correction. Everyone is human.

ABOUT THE AUTHOR

Dr. Chris McMullen has over 20 years of experience teaching university physics in California, Oklahoma, Pennsylvania, and Louisiana. Dr. McMullen is also an author of math and science workbooks. Whether in the classroom or as a writer, Dr. McMullen loves sharing knowledge and the art of motivating and engaging students.

The author earned his Ph.D. in phenomenological high-energy physics (particle physics) from Oklahoma State University in 2002. Originally from California, Chris McMullen earned his Master's degree from California State University, Northridge, where his thesis was in the field of electron spin resonance.

As a physics teacher, Dr. McMullen observed that many students lack fluency in fundamental math skills. In an effort to help students of all ages and levels master basic math skills, he published a series of math workbooks on arithmetic, fractions, long division, word problems, algebra, geometry, trigonometry, logarithms, and calculus entitled *Improve Your Math Fluency*. Dr. McMullen has also published a variety of science books, including astronomy, chemistry, and physics workbooks.

Author, Chris McMullen, Ph.D.

MATH

This series of math workbooks is geared toward practicing essential math skills:

- Prealgebra
- Algebra
- Geometry
- Trigonometry
- Logarithms and exponentials
- Calculus
- Fractions, decimals, and percentages
- Long division
- Arithmetic
- Word problems
- Roman numerals
- The four-color theorem and basic graph theory

www.improveyourmathfluency.com

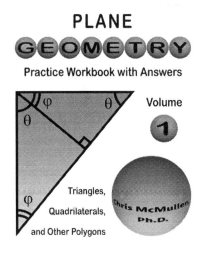

PUZZLES

The author of this book, Chris McMullen, enjoys solving puzzles. His favorite puzzle is Kakuro (kind of like a cross between crossword puzzles and Sudoku). He once taught a three-week summer course on puzzles. If you enjoy mathematical pattern puzzles, you might appreciate:

300+ Mathematical Pattern Puzzles

Number Pattern Recognition & Reasoning
- Pattern recognition
- Visual discrimination
- Analytical skills
- Logic and reasoning
- Analogies
- Mathematics

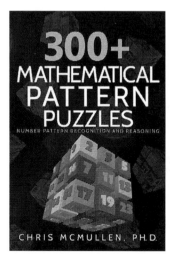

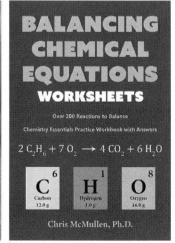

THE FOURTH DIMENSION

Are you curious about a possible fourth dimension of space?

- Explore the world of hypercubes and hyperspheres.
- Imagine living in a two-dimensional world.
- Try to understand the fourth dimension by analogy.
- Several illustrations help to try to visualize a fourth dimension of space.
- Investigate hypercube patterns.
- What would it be like to be a 4D being living in a 4D world?
- Learn about the physics of a possible four-dimensional universe.

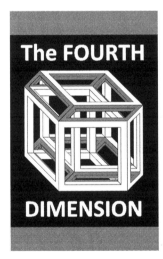

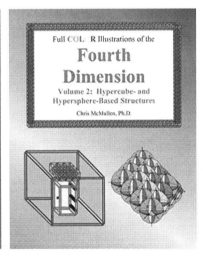

SCIENCE

Dr. McMullen has published a variety of **science** books, including:

- Basic astronomy concepts
- Basic chemistry concepts
- Balancing chemical reactions
- Calculus-based physics textbooks
- Calculus-based physics workbooks
- Calculus-based physics examples
- Trig-based physics workbooks
- Trig-based physics examples
- Creative physics problems
- Modern physics

www.monkeyphysicsblog.wordpress.com

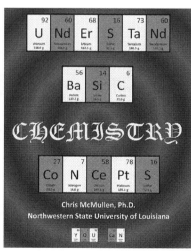

Made in the USA
Las Vegas, NV
24 September 2023

78098789R00083